Willie Ham
Teck Coal

LANDFORMING: AN ENVIRONMENTAL APPROACH TO HILLSIDE DEVELOPMENT, MINE RECLAMATION AND WATERSHED RESTORATION

Horst J. Schor
Principal
H. J. Schor Consulting

Donald H. Gray, PhD
Professor (Emeritus) of Civil and Environmental Engineering
The University of Michigan

BICENTENNIAL
1807
WILEY
2007
BICENTENNIAL

JOHN WILEY & SONS, INC.

Cover photo depicts a landformed valley fill in the Hollywood Hills filled with 350,000 cubic yards of excavated soils while still retaining the original canyon shape—Design and photo by Horst J. Schor. For details see Chapter 12—case studies.

This book is printed on acid-free paper. ⊗

Copyright © 2007 by John Wiley & Sons, Inc. All rights reserved

Published by John Wiley & Sons, Inc., Hoboken, New Jersey
Published simultaneously in Canada

Wiley Bicentennial Logo: Richard J. Pacifico

For general information about our other products and services, please contact our Customer Care Department within the United States at (800) 762-2974, outside the United States at (317) 572-3993 or fax (317) 572-4002.

Wiley also publishes its books in a variety of electronic formats. Some content that appears in print may not be available in electronic books. For more information about Wiley products, visit our web site at www.wiley.com.

Library of Congress Cataloging-in-Publication Data:
Schor, Horst J., 1938–
    Landforming: an environmental approach to hillside development, mine reclamation and watershed restoration/Horst J. Schor, Donald H. Gray.
        p. cm.
    Includes index.
    ISBN: 978-0-471-72179-6 (cloth)
    1. Soil stabilization.   2. Grading (Earthwork)   3. Soil conservation.   4. Soil-binding plants.   5. Landscape construction.   I. Gray, Donald H.   II. Title.
    TA749.S364 2007
    624.1′52—dc22

                                                                                    2006038734

Printed in the United States of America

10 9 8 7 6 5 4 3 2 1

TABLE OF CONTENTS

PREFACE

We now have the ability to alter and shape the earth's surface in profound ways. Hillside housing developments, the construction of transportation corridors, the extraction of mineral resources, and the placement of earthen landfills, all radically affect our landscape. These activities require large-scale earthmoving and grading operations, using big and powerful machines. The net result is that humans have become an important geomorphic agent; consequently, a large percentage of the earth's landforms are now anthropogenic in origin.

For the most part, these anthropogenic landforms bear little resemblance to natural landforms. Instead of the compound shapes and variable slope gradients characteristic of natural landforms, we have imposed angular shapes and planar faces with unvarying slope gradients on anthropogenic landforms. Most of us would agree that natural slopes and landforms are visually more attractive, but are they also more functional and stable? This guidebook was written (1) to show that most natural slopes and landforms are indeed functional and (2) to describe a way to achieve the dual goal of beauty and functionality using a design and grading approach known as *Landforming or Landform Grading*.

Landform Grading attempts to mimic stable natural hill slopes. Vegetation is applied on these graded slopes in a manner that is compatible with slope hydrogeology. Landform grading provides a cost-effective, attractive, and environmentally compatible way to construct slopes and landforms that are superior in the long run in terms of resistance to surficial erosion and mass wasting. Initial chapters in the book deal with the nature of surficial and mass erosion, the influence of vegetation on hillside stability, and the influence of topography on slope stability and hydrology. The geomorphic evolution of slopes is also examined with a particular view to which slope forms, ultimately, are the most enduring and stable. These initial chapters establish both the empirical and theoretical basis for the landform grading approach.

The principles of *Landform Grading* and elements of this grading approach are addressed in the chapters that follow, so, too, is the actual implementation of a landform grading plan and the public and regulatory response to such plans. This guidebook represents the distillation of thirty years of experience of the senior author who has developed landform grading designs for many projects. These projects have ranged from large hillside housing developments to surface-mining and watershed-reclamation projects to mass-grading operations. These applications are described in detail in the case studies of the last two chapters of the book.

Landforming is intended primarily as a reference handbook for practicing professionals. Information in the guidebook should prove of value to practitioners in such diverse fields as land planning, civil engineering and surveying, ecology, geotechnical engineering, mining engineering, geology, environmental horticulture, landscape architecture, and a variety of regulatory agencies (i.e., the U.S. Environmental Protection Agency [U.S. EPA], the Office of Surface Mining [OSM], and the Abandoned Mine Lands program [AML]) and environmental interest groups. Although oriented toward professional practice, it is written for students, laypersons, and other interested parties. Analytical, or technical material, in some of the chapters can be skimmed over without loss of continuity or utility. Lastly, the guidebook can be used as a reference text in college-level courses, extension courses, and workshops whose course content includes such topics as erosion control, slope stability, grading designs, mining reclamation, and watershed restoration.

We would like to acknowledge the assistance of the following persons who helped in the preparation and review of the book. Jim Holloway, community development director, City of San Clemente, was instrumental in the adoption of landform grading in the Community of Talega, California. Steven Cole, project manager, Special Project Group, City of Los Angeles Department of Water and Power, provided statistical background information and performance data for all landform-fill sites at Hollywood Hills, California. Tom Morris, senior environmental specialist, Water Quality division, Navajo EPA, advocated an environmentally responsive landforming alternative and provided continued access to measure performance at the Asaayi Lake Watershed Restoration Project. In particular the principal author would like to recognize William C. Devitt, former city engineer of the City of Anaheim, California who was the first public official to recognize the benefits of and approve the implementation of the new concept in the development of hillsides, and Dr. Hugh Marley, geotechnical engineering consultant, who validated the geotechnical soundness of landforming applications in hillside development and helped to allay initial concerns by regulatory agencies. Marie Keyes, senior engineer, Environmental Affairs and Regulatory Services, Syncrude Canada, provided past and current information about landforming performance in the Oil Sands mining and reclamation efforts. Dr. Radislav Michalowsky, Department of Civil and Environmental Engineering, University of Michigan, provided critical input and review of the theoretical underpinnings in Chapter 4. Dr. William Dietrich, Professor of Earth Sciences, University of California at Berkeley, played a key role in the development of many of the concepts presented in Chapters 2 and 5. Patricia Schor assisted in organizing the manuscript and illustrations.

HORST J. SCHOR
DONALD H. GRAY

October 2006

1 Introduction to Landform Grading and Revegetation

A great number of picture books have been compiled to show people that nature is beautiful. But the type of beauty stressed in those books is, in my opinion, the superficial kind of beauty of form evaluated solely as ornament without consideration of function and purpose. Nature is never beautiful in this sense. If things in nature are beautiful, their beauty is not superficial but the resultant form of definite purpose. In the main nature is practical—much more so than man. Its forms are functional forms derived from necessity. And precisely because in the best sense of the word they are functional, these forms are beautiful.

Andreas Feininger, *The Anatomy of Nature* (1956)

1.1 FORM AND FUNCTION IN NATURE

Performance, efficiency, and functionality are generally regarded as important goals or aspects of engineering or physical design. These are goals that tend to have well understood metrics and criteria. What about the role of beauty, aesthetics, and visual impact in design? Are these merely secondary considerations of much less importance? How can they be factored into a cost-benefit analysis, a performance evaluation, or an energy-efficiency audit? Are they considerations that a design engineer or even an earthwork contractor should be concerned about?

It would be a mistake, however, to disregard these more abstract goals in design. Humans have displayed an ageless desire for beauty that transcends simple functionality, as evidenced by our arts, crafts, architecture, and by many of our engineering structures. Greek vases, the cathedral at Chartres in France, and the Golden Gate Bridge in California are all expressions of this impulse.

Perhaps there is greater congruence between beauty and functionality than at first meets the eye. Suppose we substitute for the word "beauty" the word "form," which is an attribute or component of beauty. Form is much less subjective and more amenable to useful description. Form is also a critical component or aspect of the natural world. Form shows up everywhere in nature . . . in organic structures—whether flora or fauna. Form also shows up in nonorganic entities, ranging from mega structures, such as glaciated

1

landscapes to fourth order, glaciated landforms—such as eskers, drumlins, and moraines.

Most people would agree that natural forms are attractive and beautiful. The question is why? In the absence of some supernatural force or directive, why should nature care about beauty? In fact, nature seems to be quite ruthless; forms that are not efficient and essential for survival tend to be discarded. Evolution works to optimize efficient design and functionality. We have a great deal to learn from nature in this regard.

The intimate connection between form and function in nature is discussed at length by Feininger (1956, 1976), who describes multiple examples from the natural world—both animate and inanimate. Superior natural forms exhibit certain intrinsic properties such as clarity of organization, economy of material, symmetry of shape, and perfection of execution among others. Feininger maintains that everything in nature is designed for a purpose and that nature achieves aesthetically pleasing designs in the process. In other words, beauty is intrinsic to the very purposefulness of design in the natural world.

The concept of form following function is clearly manifest in the case of geomorphic forms. Consider the evolution of streams and upland slopes. Streams are required to transport both water and sediment. Their equilibrium profiles tend toward concave shapes over time in order to achieve this purpose as efficiently as possible; that is, gradients are steeper in the headwater region and flatten out gradually toward the mouth. Their plan forms may be sinuous or braided, depending upon the gradient and flow (discharge) at any particular point.

Slopes, likewise, transport sediment and water; in so doing, they tend toward equilibrium profiles over time. The processes in this case are more complex. Terrestrial landscapes and landforms consisting of hills and upland slopes (including valley sides) are acted upon primarily by "diffusive" and "fluvial" processes, respectively. Diffusive processes include slope wash and creep. Fluvial processes, on the other hand, are characterized by pronounced incision and formation of channels—e.g., gullying and stream-channel erosion. These processes and the resulting landform shapes are discussed in greater detail in Chapters 4 and 5. The important point to observe in the case of either stream or slope development is the presence of curvilinear shapes, compound slope forms, and general absence of planar, unvarying slope gradients.

Finally, it is important to note that beauty as a design component can be considered a "value-added" type that can provide economic as well as aesthetic benefits. This value-added component may allow easier regulatory approval, higher market value, lower maintenance and repair costs, and greater client satisfaction. It should be no surprise, therefore, that this book is titled *Landforming,* which attempts to replicate stable, natural landforms and by association their inherent beauty.

1.2   HUMAN IMPACT ON LANDFORMS

Humans have modified the surface of the earth for centuries, extracting minerals, for agricultural purposes and for urban development. In the process of this alteration, artificial landforms have been created that often bear little resemblance to natural landforms and topography. Haigh (1978) claims that humans have become an important geomorphic agent and that a large percentage of the earth's landforms are man-made and artificial (anthropogenic).

This landform alteration, or reshaping process, has largely been conceived by what might be called the "linear perspective." This perspective tends to substitute natural landforms, which are characterized by complex shapes, with much simpler landforms, characterized by planar surfaces with single, unvarying gradients. The "linear perspective," and the grading practices that derive from it, are driven to some extent by economic factors and expediency. The long-term stability and environmental impact of such grading practices have generally not been taken into account.

The prevalence of the linear perspective in conventional grading practice is somewhat puzzling. Most people would probably agree that natural landforms are more interesting and pleasing to behold. And yet those in charge of promulgating and promoting modern grading designs have apparently not been troubled by the incongruence and dissonance between natural and most artificial landforms. Numerous geomorphic studies of natural landscapes (Hack and Goodlett, 1960; Howard, 1988; Roering et al., 1999) have shown, for example, that many soil-mantled hillslopes have compound, curvilinear shapes. Some of these hillsides are not only convex in profile but also in planform. Parsons (1988) recognized that slope units may be planar, concave, or convex in plan, just as they may be in profile. Accordingly, nine possible slope-unit shapes are required for completeness, as shown schematically in Figure 1.1. Where slopes transition into valley networks or convergent parts of the landscape, slope and channel profiles tend to become concave. Studies by Hancock et al. (2003), for example, have shown that soil-mantled, fluvial erosion–dominated catchments generally have convex upper-hillslope profiles with concave profiles developing further downslope, as shown in Figure 1.2.

As drainage areas increase in these channelized or incised portions of the landscape, slope gradients tend to decrease, thereby leading to concave slope profiles. Apparently, there has been a general failure to recognize the existence of these more complex slope forms and to realize that rectilinear profiles and planar slopes are seldom found in nature.

What is the long-term stability of artificial, planar-slope shapes versus more complex slope forms that include concave- and convex-slope profiles? This question is examined in some detail in Chapters 4 and 5. Even in land restoration and reclamation work, there has been a tendency to use artificial landforms with rigidly conceived slope forms and profiles.

PROFILE FORM

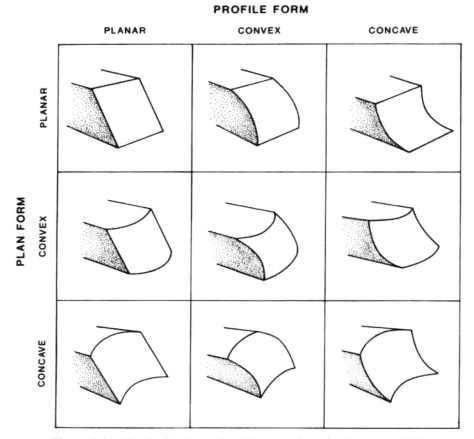

Figure 1.1   Nine basic shapes for hillslope units (after Parsons, 1988).

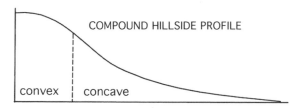

Figure 1.2   Idealized cross section of a natural hillslope in a soil-mantled landscape.

One could ask why it has not occurred more often to persons in charge of these restoration efforts to utilize at least some natural landform shapes? Why have not more owners and regulatory agencies considered the long-term environmental and aesthetical impact of such artificial reshaping and remolding of natural topography upon future generations? Landforming techniques described in this book provide a basis for adopting a new land restoration and reclamation paradigm.

1.3   HISTORICAL DEVELOPMENT

Earlier urban development generally occurred on mostly level land that was fairly easy to build upon. Over time, more towns and cities were built in areas with greater topographic relief. Development of towns and cities in hilly terrain was feasible, steepness of slopes notwithstanding, if located on dense and stable bedrock such as igneous or metamorphic rocks.

One way of avoiding incompatibility between a proposed land use and the underlying terrain is to adopt a landscape-planning approach that is based on ecological rather than purely economical considerations. This entails identifying suitable land uses based on topographic, geologic, hydrologic, pedalogic, and botanic factors. One of the primary and most forceful exponents of this "design with nature" approach to landscape planning was Ian McHarg (1969), whose book had a seminal impact on the field of land planning and landscape architecture. Other exponents of the design-with-nature approach followed in his footsteps in an attempt to integrate land planning, land science (geology, geomorphology, and geography), and landscape design. The importance of slopes and topography in land-use planning has been emphasized by Marsh (2005). He noted that land uses have slope limitations and showed how slopes have been misused in modern land developments.

The impact of urban development in hilly terrain on the natural topography could normally be minimized if low densities were maintained, because building sites could be fitted into the existing terrain with minimum grading and access-road widths. Alignment and grades were flexible enough to adjust to natural conditions. Under these circumstances, geotechnical concerns, such as slope stability and bearing capacity, could be handled with small scale remediation as opposed to massive grading and earthwork. This resulted in urban development that fitted or blended into the landscape with minimal disturbance and earth movement as shown in Figure 1.3.

An entirely different situation is apt to occur when mining, landfilling operations, and intensive urbanization move into hilly terrain. Potential problems are compounded when the underlying bedrock is sedimentary and when major geotechnical instability problems have to be considered, such as faults, landslides, groundwater, compressible (or expansive) soils, buried boulders, and so forth. Under these circumstances, single-family, detached-housing lots, building pads for multiple, attached-family units, and pads for commercial,

Figure 1.3 Illustration of hillside development that blends or fits into the terrain without excessive topographic disturbance.

industrial, and institutional buildings generally require large-scale grading, landform alteration, and remedial treatments to create large, flat, and level building sites.

Such use also calls for a more extensive circulation system designed for wider roads to accommodate greater traffic volumes, larger horizontal and vertical radius street curves, and flatter grades for higher and safer speeds. Other infrastructure facilities have their own special location and site needs, that is, reservoirs, pump stations, waste-disposal landfills, water-treatment plants, gravity sewers, and so forth. When these land uses are combined with high relief and adverse geologic or soil conditions, the results often require extensive grading and reshaping of the natural topography with the objective of (1) creating level building pads and (2) mitigating or correcting geotechnical instability problems.

Over the years, stringent design standards have been established by regulatory agencies and the civil engineering profession to meet these objectives. The primary emphasis has been on meeting short-term stability requirements and runoff control in grading designs. Geotechnical slope-stability analyses (Abramson et al., 2002) seldom if ever included time as an explicit variable. The result was a visual product of flat surfaces and rigid, linear, and angular slope forms with little resemblance to the original natural landscape. This also tended to result in a man-made environment with few redeeming aesthetic or visual qualities. The fundamentals of conventional grading practice are treated in greater detail in Chapter 6.

Landform grading concepts were developed to redress these deficiencies and to introduce aesthetic considerations into hillside develop-ment. Early work examined various elements of such projects to determine which would be best suited for possible rethinking and reconfiguration. These early efforts led to the realization that hillside grading transformed natural topographic elements (swales, ridge lines, and side slopes) into two basic components, namely, flat pad areas and slopes, as shown in Figure 1.4.

It also became apparent that the pad areas quickly became obscured by structures, roads, and other appurtenant development features. On the other hand, the slope component continued to stand out as a permanent visual element for better or worse, as illustrated in Figure 1.4.

Accordingly, initial studies focused on the slope element. This element was compared to equivalent natural slope forms to determine if Nature could provide some useful lessons and directions with regard to reintroducing the functional beauty of a natural hillside into mass-graded, man-made environments. These initial studies led to the discovery that the shape of the slope element had a significant influence on aesthetic appearance. Furthermore, slope shape

Figure 1.4 Transformation of natural hillside topography as a result of modern grading practices into two basic components, namely, flat pad areas and planar slopes.

and form also impacted the configuration of building pads above and below the slope and, ultimately, on road alignments and the configuration and placement of structures. Important characteristics and attributes of the slope element are considered in greater detail in Chapters 2, 4, and 5.

Follow-on work consisted of careful visual observations and photographic studies of natural hillside slopes throughout the world. Their morphology was measured on topographic maps to determine their size, shape, and exact proportions. The map studies provided additional information about scale and proportion. The finding that emerged from this study was the recognition that natural hillsides consisted basically of a series of universal slope "building blocks," or components, which tended to repeat themselves regardless of the local soils and climatic conditions.

These hillside components consisted in their general form in a series and variety of concave, convex, and occasionally linear elements. Some occurred in relatively simple arrangements while others occurred in more complex arrays. All were ultimately the product or the result of erosional processes. Additional information and attributes about these slope forms and arrays are provided in Chapter 8.

Landform grading essentially attempts to: (1) respect the underlying, basic landforms by preserving or replicating them and their associated vegetative patterns and (2) re-create or mimic the important, stable natural hillsides with their rich variety of different slope elements and forms. When this conceptual approach is applied to hillside housing developments a very different topography and configuration of building pads, roads, and drainage ways emerges, as shown schematically in Figure 1.5. A photograph of an actual hillside development where landform grading was employed is shown in Figure 1.6.

Hillside grading fundamentals are discussed briefly in Chapter 6. The main characteristics and differences between landform and conventional grading practices are discussed and compared in Chapter 7. The use of *landforming* techniques to repair and rejuvenate man-made or damaged natural landscapes are discussed in this chapter as well. Natural slope elements and forms are identified and described in Chapter 8. Different approaches to grading a natural landform with distinct topographic features are presented herein as a prelude to the chapters that follow.

Aerial photos of natural landforms, as well as the different approaches to grading them, are presented in Figures 1.7–1.8. The underlying topography in the first case is characterized by an east-west trending, primary ridge line, from which a series of secondary ridges and valleys descend in a north-south trending direction. A conventional approach of cutting perpendicularly across the secondary features with little or no concern to preserving or replicating the original topography is illustrated in Figure 1.7. This approach ultimately results in a terraced, stair-stepped landscape that obliterates the features of the original terrain.

Figure 1.8 shows a similar approach, however, the underlying landform was a broad, round slope feature. The grading in this case created terraced

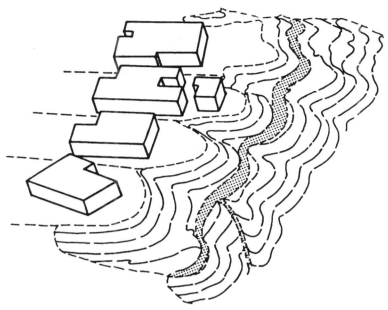

Figure 1.5   Schematic illustration of landform grading approach showing topography of slopes, configuration of building pads and position of drainages.

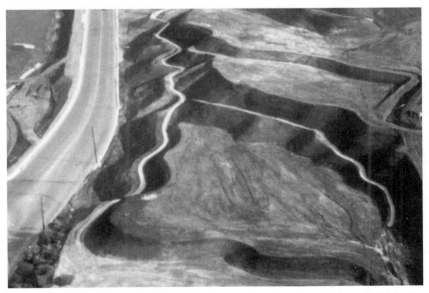

Figure 1.6   Photograph illustrating landform grading approach in a hillside development project in California.

Figure 1.7   Conventional grading that cuts across secondary ridges and valleys, resulting in a terraced, stair-stepped landscape that obliterates the original terrain features.

slopes and building pads but, nevertheless, replicated the underlying landform. This approach displays aspects of both conventional grading practice and landform grading. Once fully developed this grading approach still retains the character of the original, underlying terrain.

1.4   OBJECTIVES AND CHALLENGES

There are some fairly formidable challenges that stand in the way of more widespread adoption of landform grading practice. One of the main purposes of this book is educational, namely, to present in a single source and in a

Figure 1.8 Alternative grading approach on a broad, round slope feature. Grading created terraced slopes and building pads but nevertheless replicated the underlying landform.

concise form all the available information about *landforming* and its attributes. Another purpose is to make the case for *landforming* by demonstrating its advantages by comparing the relative stability of simple, planar slopes versus more complex slope shapes and by presenting case studies of actual projects.

The following are some of the challenges to more widespread adoption and implementation of landform grading practice:

1. Overcoming the inertia of the civil engineering profession both in the office design and field surveying departments that have relied on traditional grading designs and simple landforms.

2. Providing more direction, training, and control to the planning professions. These include land planners, landscape architects, and terrestrial ecologists.

3. Informing geotechnical engineers about the merits of using more complex slope shapes. Geotechnical engineers normally work with and analyze only the stability of planar slopes with linear profiles.

4. Overcoming the skepticism about and reluctance to approve new grading approaches on the part of regulatory agencies.

Landform Grading is applicable not only to hillside housing developments but also to land-reclamation and watershed-restoration work following mining operations. In this case, the slope forms that are created are not simply cut slopes but artificial embankments, ridges, and depressions. The same principles still apply, namely, creating stable landforms that are visually compatible with the surrounding natural landscape and in harmony with regional vegetation patterns and surface hydrology.

1.5   REFERENCES

Abramson, L. W., T. S. Lee, S. Sharma, and G. M. Boyce. 2002. *Slope Stability and Stabilization Methods.* 2nd ed. New York: John Wiley & Sons.

Feininger, Andreas. 1956. *The Anatomy of Nature: How Function Shapes the Form and Design of Animate and Inanimate Structures Throughout the Universe.* New York: Crown Publishers.

———. 1976. *Forms of Nature and Life.* New York: Viking Press.

Hack, J. T., and J. C. Goodlett. 1960. Geomorphology and forest ecology of a mountain region in the central Appalachians. *U.S. Geological Survey Professional Paper* 347, 1–66.

Haigh, M. J. 1978. *Evolution of Slopes on Artificial Landforms, Blaenavon, United Kingdom.* Research Paper No. 182. University of Chicago, Dept. of Geography, Chicago, IL.

Hancock, G. R., R. Loch, and G. R. Willgoose. 2003. The design of postmining landscapes using geomorphic guidelines. *Earth Surface Processes and Landforms* 28:1097–1110.

Howard, A. D. 1988. Equilibrium models in geomorphology. In *Modelling Geomorphological Systems,* ed. M. G. Anderson, 49–72. New York: John Wiley & Sons.

Marsh, W. M. 2005. *Landscape Planning: Environmental Applications.* 4th ed. New York: John Wiley & Sons.

McHarg, I. L. 1969. *Design with Nature.* New York: John Wiley & Sons.

Parsons, A. J. 1988. *Hillslope Form.* London: Routledge.

Roering, J. J., J. W. Kirchner, and W. E. Dietrich. 1999. Evidence for nonlinear, diffusive sediment transport on hillslopes and implications for landscape morphology. *Water Resources Research* 35 (3):853–70.

2 Surficial Erosion and Mass Wasting of Slopes

2.1 INTRODUCTION

Hills and uplands form as the result of tectonic forces that warp the earth's crust. Plutonic rock masses can also push up through the crust, forming mountain ranges such as the Sierra Nevada. Volcanic activity may bring up molten rock from the earth's interior and deposit it on the surface to create mountains and volcanoes as well. These mountain areas and uplands, in turn, can be degraded and worn down by the twin processes of surficial erosion and mass wasting. (The role of these geologic processes in shaping the earth's landscape as well as approaches to slope evolution are discussed further in Chapter 5.)

To control or prevent this wearing or wasting away of the earth's surface, it is first necessary to understand these two processes of degradation and the factors that control them. Man-made or manufactured slopes—for example, road cuts and embankment slopes—are subject to the same degradation processes. While these two processes share many similarities, they also differ in important respects. Surficial erosion entails the detachment and transport of individual particles whereas mass wasting entails the movement of relatively large, initially intact masses of soil and/or rock along critical failure surfaces. Gravity is the main driving force behind mass wasting, and wind and running water are the principal agents of surficial erosion. The role and function of vegetation also differ substantially between these two processes.

2.2 DEFINITIONS

2.2.1 Surficial Erosion

Surficial erosion, or soil erosion, is the removal of surface layers of soil by the agencies of wind, water, and ice. Soil erosion involves a process of both particle *detachment* and *transport* by these agencies. Erosion is initiated by drag, impact, or tractive forces acting on individual particles of soil at the surface. Weathering processes, such as frost action and wet-dry cycling, set the stage for erosion by breaking up rock into smaller particles and weakening bonds between particles.

Rainfall and wind erosion are the two most common types of erosion. Rainfall erosion starts with falling raindrops. When these drops impact on bare or fallow ground, they can dislodge and move soil particles a surprising distance. At the onset of runoff, water collects into small rivulets, or small channels called rills, that can erode. These rills may eventually coalesce into larger and deeper channels called gullies. Gullying is a complex and destructive process that once started is difficult to stop. Bare or unprotected earth surfaces are the most vulnerable to all forms of surficial erosion.

2.2.2   Mass Wasting

Mass wasting is a descriptive name for the downward and outward movement of slope-forming materials—natural rock, soils, artificial fills, or combinations of these materials. The terms *mass movement* or *mass erosion* are sometimes used. *Mass movements* are popularly known as landslides. Strictly speaking, however, landslides, or slides, refer to a particular type of mass wasting.

Unlike superficial erosion, mass wasting involves the sliding, toppling, falling, or spreading of fairly large and sometimes relatively intact masses. Mass movements have been classified into categories based largely on the type of movement and material involved (Varnes, 1978). A slide is a relatively slow slope movement in which a shear failure occurs along a specific surface or combination of surfaces in a slope.

2.2.3   Salient Characteristics and Differences

Many of the same slope, soil, and hydrologic factors that control surficial erosion also control mass wasting (e.g., steepness of slope and shear strength of soil). The two processes differ, however, in some important respects. The salient characteristics of surficial erosion on the one hand and mass wasting on the other are contrasted in Figure 2.1.

Precipitation, a key factor directly affecting rainfall erosion, only affects mass wasting indirectly, via its influence on the groundwater regime at a site. In contrast, geologic conditions such as orientation of joints and bedding planes in a slope can have a profound influence on mass stability but not on surficial erosion. Vegetation has an important influence on both erosion and shallow mass movement.

Different predictive techniques or models have been developed to determine soil losses from surficial erosion or the likelihood of catastrophic slope failure in the case of mass wasting. Rainfall and wind erosion, for example, are controlled by a number of soil, climatic, and topographic factors, including intensity and duration of precipitation, ground roughness, length and steepness of slope, inherent soil erodibility, and type or extent of cover. All of these factors are taken into account explicitly in the Universal Soil Loss Equation (Wischmeier and Smith, 1978).

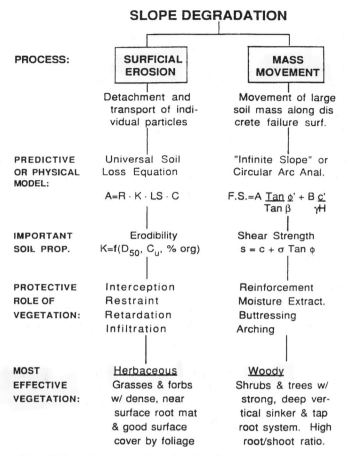

Figure 2.1   Salient characteristics of surficial erosion vs. mass movement.

2.3   NATURE OF SURFICIAL EROSION

2.3.1   Agents and Types of Erosion

The primary agents of erosion include water, wind, and ice. These agents can scour and remove soil particles as a result of water (or air) flowing past, impacting upon, or exiting from the surface of a soil. Each agent can also erode soil in quite different manners. Rainfall erosion, for example, can occur in the form of splash, sheet, or rill erosion.

Degradation processes often act jointly or in combination with one another. Glaciated landscapes, for example, may be the product of glacio-fluvial processes.

2.3.2   Mechanics of Erosion

Prevention and control of erosion depends on understanding the mechanics of the erosion process. Erosion is basically a two-fold process that involves (1) particle detachment and (2) particle transportation.

The forces acting on particles near a fluid-bed boundary, or interface, are shown schematically in Figure 2.2. Drag or tractive forces exerted by the flowing fluid are resisted by inertial or cohesive forces between particles.

Erosion protection essentially consists of (1) decreasing drag or tractive forces by decreasing the velocity of water flowing over the surface or by dissipating the energy of the water in a defended area and (2) increasing resistance to erosion by protecting and reinforcing the surface with a suitable cover (e.g., with vegetation) or by increasing interparticle bond strength (e.g., with cementing agents).

2.4   PRINCIPAL DETERMINANTS OF EROSION

2.4.1   Rainfall Erosion

Rainfall erosion is controlled by four basic factors, namely, climate, soil type, topography, and vegetative cover. This relationship can be expressed schematically as follows:

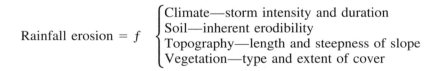

$$\text{Rainfall erosion} = f \begin{cases} \text{Climate—storm intensity and duration} \\ \text{Soil—inherent erodibility} \\ \text{Topography—length and steepness of slope} \\ \text{Vegetation—type and extent of cover} \end{cases}$$

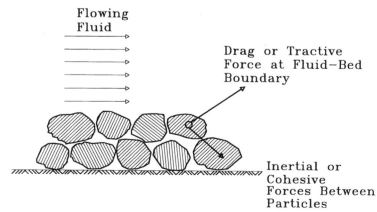

Figure 2.2   Schematic diagram—of forces acting on particles at fluid-bed boundary.

The most important climatic parameters controlling rainfall erosion are intensity and duration of precipitation. Wischmeier and Smith (1978) have shown that the most important single measure of the erosion-producing power of a rainstorm is the product of rainfall energy times the maximum 30-minute rainfall intensity. Raindrops impacting on bare soil not only cause erosion but also tend to compact the soil and decrease infiltration capacity.

The susceptibility of a soil to erosion is known as its "erodibility." Some soils (e.g., silts and silty sands) are inherently more erodible than others (e.g., coarse, well-graded gravels). In general, increasing the organic content and clay-size fraction of a soil decreases erodibility. Erodibility also depends upon such parameters as soil texture, antecedent moisture content, void ratio, exchange ions, pH, and composition or ionic strength of the eroding water.

There is no simple and universally accepted erodibility index for soils. Instead, various tests have been proposed for this purpose, including the Soil Conservation Service (SCS) dispersion test (Volk, 1973), crumb test (Emerson, 1967), and pinhole test (Sherard et al., 1976). A suggested hierarchy of erodibility based on the Unified Soil classification system is:

Most Erodible → Least Erodible
ML > SM > SC > MH > OL >> CL > CH > GM > SW > GP > GW

This erodibility hierarchy is simple but based on gradation and plasticity indices of remolded or disturbed soils. Accordingly, it fails to take into account the effects of soil structure, void ratio, and antecedent moisture content. Wischmeier et al. (1971) published an erodibility nomograph for use with the Universal Soil Loss Equation (see Section 2.6), which is based on easily measured soil properties.

Topographic variables influencing rainfall erosion are (a) slope angle, (b) slope length, and (c) size and shape of watershed. The effect of slope profile or shape is discussed in detail in Chapter 4 because this variable is a key consideration in landform grading. The influence or importance of length tends to increase as slopes become steeper. For instance, a doubling of slope length from 100 to 200 feet will only increase soil losses by 29 percent on a 6-percent slope, whereas the same doubling of slope length in a 20-percent slope will result in a 49-percent increase in soil loss. This is one of the reasons for benching or terracing long, steep slopes.

Vegetation plays an extremely important role in controlling rainfall erosion. Removal or stripping of vegetation by either human or natural agencies (e.g., wildfires) often results in accelerated erosion. Conversely, any measure that either preserves or enhances vegetation establishment will significantly retard soil loss and minimize erosion. The protective mechanisms of vegetation in preventing both surficial erosion and shallow mass wasting are discussed at length in Chapter 3.

2.4.2   Wind Erosion

Wind erosion is controlled by the same basic factors that control rainfall erosion. Unlike rainfall erosion, topographic parameters such as length and steepness of slope are relatively unimportant in the case of wind erosion. On the other hand, surface roughness and the presence of low barriers that act as wind breaks and sediment traps can be important.

Because wind erosion is little influenced by topography compared to rainfall erosion, no further discussion of the characteristics and mechanics of wind erosion is presented in this book, which focuses on grading practice to purposely shape slopes and alter topography to minimize erosion. Further information on the physics of wind erosion and its control are presented elsewhere (Bagnold, 1941; Chepil and Woodruff, 1963).

2.5   TYPES OF WATER EROSION

Water erosion manifests itself in many different ways. Rainfall erosion begins with raindrop splash and can progress ultimately to gullying and stream-channel erosion. The salient characteristics of different types of water erosion are briefly reviewed below.

Raindrop Splash: Raindrop splash results from the impact of water drops falling directly on exposed soil particles or thin water surfaces covering the ground. Tremendous quantities of soil can be splashed into the air in this manner. On bare ground, it has been estimated (Ellison, 1948) that as much as 100 tons per acre can be splashed into the air in a heavy storm. Splashed particles may move more than 2 feet vertically and 5 feet laterally on level ground. On steep slopes this splashing will cause a net downslope movement of soil.

Sheet Erosion: Sheet erosion is the removal of soil from sloping land in thin layers or sheets. From an energy standpoint, raindrop erosion appears to be more important than sheet erosion, because most raindrops have velocities of about 20 to 30 feet per second (fps), whereas overland flow velocities are about 1 to 2 fps. Dry ravel and slope wash are forms of sheet erosion; the former occurs when surface layers of coarse-textured soil dry out and lose their apparent cohesion while the latter occurs when rainfall erodes without causing rilling or gullying. Sheetlike erosion is an important mechanism of slope retreat and source of sediment in cut slopes in granitic and andesitic soils. Highway cuts in these soils often give the impression of stability (e.g., rills and gullies are absent) yet discharge tons of soil year after year into roadside ditches.

Rill Erosion: Rilling is the removal of soil by water from very small but well-defined, visible channels, or streamlets, where there is concentration of overland flow. An example of rill erosion at an urban construction site is shown in Figure 2.3. Rilling generally is more serious than sheet erosion because runoff velocities are higher in the rills, or channels. Most rainfall erosion losses occur because of rilling (Schwab et al., 1966).

Rill erosion is most serious where intense storms occur in watersheds or sites with high runoff-producing characteristics and loose, shallow topsoil. Rills are sufficiently large and stable enough to be seen readily, but they are also small enough to be removed easily by normal tillage and grading operations.

Gullying: Gullies are intermittent stream channels larger than rills. These channels carry water during and immediately after rains, and, unlike rills, gullies cannot be obliterated by normal tillage. Various statistical models for predicting gully growth and development have been proposed by Beer and Johnson (1963). Gullies tend to form where large volumes of runoff are concentrated and discharged onto steep slopes with erodible soils, for example, undefended culvert outlets. Gullying is common in grass lands and likely the principal form of erosion in steep, forested watersheds. The conditions under which gullies are most likely to occur are discussed further in Section 2.10, which deals with mass-stability and channel-erosion thresholds in watersheds.

Gullies may not be as significant as rills in terms of total quantities of soil eroded, but they can be quite destructive in terms of damage to roadways, embankments, and watersheds, as shown in Figure 2.4. They are also difficult to control and arrest. Effective gully control must stabilize both the channel bottom and head cuts. Continued down-cutting

Figure 2.3   Rill erosion at an urban construction site, Ann Arbor, Michigan.

Figure 2.4 Hillside gully erosion caused by uncontrolled and concentrated discharge of runoff from parking lot.

of gully bottoms leads to deepening and widening, whereas headcutting extends the channel into ungullied headwater areas, and increases the stream net and its density by developing tributaries.

Stream Channel Erosion: Stream-channel erosion consists of soil removal from stream banks and/or sediment scoured from the channel bottom. Stream-channel erosion should be considered separately from the rainfall-associated types of erosion discussed previously. A number of hydrologic, hydraulic, and geomorphic variables govern the behavior of fluvial systems. These variables are in a dynamic equilibrium with one another. Investigators, such as Leopold (1994), have determined several general relationships among these variables that are useful in analyzing stream activity.

The causes of channel and stream-bank erosion are linked to fluvial geomorphology and the interrelationship between hydraulic and hydrologic variables. From the standpoint of fluvial geomorphology, there appear to be three main actions or events (Keown et al., 1977; Leopold, 1994) that result in stream erosion:

Widening: Channel enlargement caused by increased stream flow and/or sediment discharges.

Deepening: Scouring of the channel bottom caused by increased stream flows and or changes in slope.

Sinuosity Change: Bank loss that occurs during and upon a change in planform or stream-meander configuration. This bank loss is usually accompanied by accretion somewhere else along the affected reach.

Guidelines for selecting and implementing environmentally sensitive channel and stream-bank erosion-protection measures are provided in a comprehensive guidance manual prepared for use by highway engineers and their consultants by the National Cooperative Highway Research Program (NCHRP, 2005).

Groundwater Erosion: Groundwater erosion is the removal of soil caused by groundwater seepage or movement toward a free face. Such erosion is commonly referred to as piping. The phenomenon is also known as spring sapping—literally the detachment and movement of soil particles at the point of emergence of a spring or seep in the ground. Piping occurs when seepage forces exceed intergranular stresses or forces of cohesion.

2.6   SOIL LOSS PREDICTIONS

2.6.1   Historical Development

A semiempirical equation known as the Universal Soil Loss Equation (USLE) was developed by the U.S. Department of Agriculture's Agricultural Research Service (ARS) in the early 1960s. The Universal Soil Loss Equation was modified and adapted to different regions of the United States (USDA Soil Conservation Service, 1972,) and also for use at urban or highway construction sites (Israelson, 1980). These latter uses have stirred some controversy about the limits and applicability of the USLE and prompted numerous efforts for further revision (Renard et al., 1991).

The USLE takes into account all the factors known to affect rainfall erosion, namely, climate, soil, topography, and vegetative cover. It is based on a statistical analysis of erosion measured in the field on scores of test plots under natural and simulated rainfall. The annual soil loss from a site is predicted according to the following relationship:

$$A = R \cdot K \cdot LS \cdot C \cdot P \qquad (2.1)$$

where A = computed soil loss (e.g., tons) per acre for a given storm period or time interval;

R = rainfall factor;

K = soil erodibility value;

L = slope length factor;

S = steepness factor;

C = vegetation factor; and

P = erosion control practice factor.

2.6.2   Applications of the Universal Soil Loss Equation (USLE)

Detailed procedures and charts for estimating soil losses and calculating values for the various parameters in the USLE are given in Gray and Sotir (1982) and Goldman et al. (1986). This information is not repeated herein; instead, the objective is to show which parameters have the greatest variability and effect on soil loss and which can be easily changed.

The climatic (**R**) factor and erodibility (**K**) factor only vary within one order-of-magnitude. The cover (**C**) factor and topographic (**LS**) factor, on the other hand, can vary over several orders-of-magnitude, as shown in Table 2.1 and Table 2.2. respectively. Moreover, unlike the other factors, the cover (**C**) factor and topographic (**LS**) factor can be modified substantially to suit a particular erosion control objective.

The topographic (**LS**) factor can be modified easily by reducing slope length, that is, by converting a long, steep slope into a series of short, steep slopes. This is commonly accomplished by benching or terracing. The placement of benches at intervals of 10 feet, for example, in a 100-foot long slope would reduce the soil loss to one-third of the original value. This reduction can be observed by examining the soil loss, or **LS** factor, values in Table 2.1 that correspond to a 100-foot long versus a 10-foot long slope for a given slope gradient. The combined effect of slope length, gradient, and profile shape are examined further in Chapter 4.

The vegetative cover, or **C** factor, affects erosion via three separate and distinct but interrelated zones of influence, namely, canopy cover, ground cover (vegetative cover in direct contact with the soil), and vegetative or crop

TABLE 2.1   Topographic (LS) Factors for Slopes

Slope Ratio (H/V)	Slope Gradient 'S', %	LS Values for Selected Slope Lengths 'L', ft(m)						
		10 (3.0)	30 (9.1)	50 (15.2)	100 (30.5)	300 (91.0)	500 (152.0)	1000 (305.0)
20 to 1	5	0.17	0.29	0.38	0.53	0.93	1.20	1.69
10 to 1	10	0.43	0.75	0.97	1.37	2.37	3.06	4.33
8 to 1	12.5	0.61	1.05	1.36	1.92	3.33	4.30	6.08
6 to 1	16.7	0.96	1.67	2.15	3.04	5.27	6.80	9.62
5 to 1	20	1.29	2.23	2.88	4.08	7.06	9.12	12.90
4 to 1	25	1.86	3.23	4.16	5.89	10.20	13.17	18.63
3 to 1	33.5	2.98	5.17	6.67	9.43	16.33	21.09	29.82
2.5 to 1	40	4.00	6.93	8.95	12.65	21.91	28.29	40.01
2 to 1	50	5.64	9.76	12.60	17.82	30.87	39.85	56.36
1.75 to 1	57	6.82	11.80	15.24	21.55	37.33	48.19	68.15
1.50 to 1	66.7	8.44	14.61	18.87	26.68	46.22	59.66	84.38
1.25 to 1	80	10.55	18.28	23.60	33.38	57.81	74.63	105.55
1 to 1	100	13.36	23.14	29.87	42.24	73.17	94.46	133.59

Source: From Israelson (1980).

TABLE 2.2   Vegetation Factor (C) Values for Pasture, Rangeland, and Idle Land

Type and Height of Raised Canopy	Canopy Cover (%)	Canopy Type	C Values for Selected Canopy and Ground-Cover Conditions					
			Ground Cover (%)					
			0	20	40	60	80	95–100
No significant canopy	—	G	.45	.20	.10	.042	.013	.003
		W	.45	.24	.15	.090	.043	.011
Canopy of tall weeds or short brush (0.5 m fall height)	25	G	.36	.17	.09	.038	.012	.003
		W	.36	.20	.13	.082	.041	.011
	50	G	.26	.13	.07	.035	.012	.003
		W	.26	.16	.11	.075	.039	.011
	75	G	.17	.10	.06	.031	.011	.003
		W	.17	.12	.09	.067	.038	.011
Appreciable brush or bushes (2 m fall height)	25	G	.40	.18	.09	.040	.013	.003
		W	.40	.22	.14	.085	.042	.011
	50	G	.34	.16	.09	.038	.012	.003
		W	.34	.19	.13	.081	.041	.011
	75	G	.28	.14	.08	.036	.012	.003
		W	.28	.17	.12	.077	.040	.011
Trees but no appreciable low brush (4 m fall height)	25	G	.42	.19	.10	.041	.013	.003
		W	.42	.23	.14	.087	.042	.011
	50	G	.39	.18	.09	.040	.013	.003
		W	.39	.21	.14	.085	.042	.011
	75	G	.36	.17	.09	.039	.012	.003
		W	.36	.20	.13	.083	.014	.011

G: Cover at surface is grass, grasslike plants, decaying compacted duff, or litter at least 2 inches deep.
W: Cover at surface is mostly broadleaf, herbaceous plants with little lateral-root network near surface.

Source: USDA Soil Conservation Service (1978).

residue at or beneath the surface. The effects of these three constituent influences can be observed in Table 2.2. For completely bare or fallow ground, the **C** factor is unity.

The influence of ground cover trumps that of the canopy. This influence can be readily ascertained by comparing **C** factor values in the row versus column headings of Table 2.2. Even in the absence of appreciable canopy, the **C** factor value can drop to a value of only 0.003 when ground cover reaches 95 percent. This corresponds to almost a thousandfold reduction in erosion losses over the fallow- or bare-ground case. None of the other variables or factors are amenable to management with such dramatic results as this one. Mulches and biotechnical ground covers (e.g., nets, erosion control blankets, turf reinforcement mats, etc.) provide initial erosion protection and enhance vegetative establishment and performance. Factor C values for different types of ground covers are listed in Table 2.3.

Mulches provide immediate protection on bare soils and may perform better than temporary seedings in certain instances. Straw is particularly effective as a mulch or ground cover (Kay, 1983). Straw that has been tacked down

TABLE 2.3   C Factors for Different Ground Covers

Factor	Type of Cover	Soil Loss Reduction C %
None	00	1.00
Native vegetation (dense, undisturbed)	99	0.01
Temporary seedings:		
90% cover, annual grasses, no mulch	90	0.10
Wood fiber mulch, 0.75 ton/acre, with seed	50	0.05
Excelsior mat, jute netting (slopes up to 2:1)	70	0.30
Straw mulch		
1.5 tons/acre (3.4 t/ha), tacked down	80	0.20
4.0 tons/acre (9.0 t/ha), tacked down	95	0.05

Source: From Goldman et al. (1986).

provides better protection (lower soil loss) than wood fiber mulch at equal application rates. Factor-**C** values tend to change with time following certain types of surface treatment, such as mulching, seeding, and transplanting. For example, factor-**C** values for grass may decrease from 1.00 (for fallow, bare ground) to about 0.01 between time of initial seeding and full establishment. In this case, an average or weighted factor-**C** value can be computed for a construction period, using time as the weighting parameter and estimating **C**-factor values for successive time intervals.

2.6.3   Limitations of USLE

The USLE is an empirically based equation that predicts sheet and rill erosion for relatively small areas. Although it explicitly takes into account all the factors known to affect rainfall erosion and is widely employed around the world, it has the following limitations:

- *It is empirical.* Considerable judgment is required in assigning correct values to some of the factors in certain situations. This is particularly true in the case of the vegetation or cover factor.
- *It predicts average annual soil loss.* The rainfall factor is based on a 2-year, 6-hour rainfall. Unusual storms or weather events during a particular year could produce more sediment than predicted.
- *It does not predict gully erosion.* The USLE predicts soil loss from sheet and rill erosion—not erosion resulting from concentrated flow in large channels or gullies.
- *It does not predict sediment delivery.* The USLE predicts soil loss—not sediment deposition. Soil lost from the upper portions of a slope or watershed will not automatically end up as sediment in a body of water. Instead, it may collect on the lower portions of a slope or watershed.

In spite of their limitations, the USLE and revised versions (RUSLE) provide a simple, straightforward method of estimating soil losses, identifying critical areas, and evaluating the effectiveness of soil-loss reduction measures. They provide a rational basis for sizing sediment basins and other sediment collection systems at construction sites. Significantly, the USLE provides an idea of the range of variability of each of the parameters, their relative importance in affecting erosion, and the extent to which each can be changed or managed to limit soil losses.

2.7   EROSION CONTROL PRINCIPLES

Surficial erosion can be controlled or prevented by observing some basic principles. These principles are universally applicable; they should be observed regardless of whether conventional or soil-bioengineering treatments are contemplated. Erosion-control principles are based on common sense, but they are frequently violated in site-development work. Many erosion control measures and products have been introduced over the years; they are more effective when applied in conjunction with the principles that are enumerated below:

1. Fit the development plan to the site. Avoid extensive grading and earthwork in erosion prone areas.
2. Install hydraulic conveyance facilities to handle increased runoff.
3. Keep runoff velocities low.
4. Divert runoff away from steep slopes and denuded areas by constructing interceptor drains and berms.
5. Save native site vegetation whenever possible.
6. If vegetation must be removed, clear the site in small, workable increments. Limit the duration of exposure.
7. Protect cleared areas with mulches and temporary, fast growing herbaceous covers.
8. Construct sediment basins to prevent eroded soil or sediment from leaving the site.
9. Install erosion control measures as early as possible.
10. Inspect and maintain control measures.

Observance of these ten basic principles will greatly minimize erosion losses. Goldman et al. (1986) discuss and elaborate on the importance of these basic principles.

Some additional discussion is warranted here on the impact of grading practices on soil erosion. Both transportation corridors and residential developments in steep terrain require that some excavation and regrading be carried out to accommodate roadways or building sites. The manner in which this grading is planned and executed and the nature of the resulting topography or landforms that are created affect not only the visual or aesthetic impact of the development but also the mass and surficial stability of the slopes and effectiveness of landscaping and revegetation efforts. A planar slope with a linear profile will exhibit higher soil losses than a slope with a concave or decreasing gradient near the toe. The influence of topography and slope morphology on erosion are considered in greater detail in Chapter 4. Likewise, drainage channels brought down and across a slope in a curvilinear manner, which lengthens flow path and reduces gradient, are less susceptible to erosion than exposed channels brought directly down the face of a slope. These are fundamental principles of landform grading, and they are discussed at greater length in Chapter 7. Landform grading techniques mitigate and minimize erosion problems by observing these principles and by mimicking stable, natural slopes.

2.8   NATURE OF MASS WASTING

2.8.1   Types of Slope Movement

Various schemes have been proposed over the years for classifying and describing slope movements (Varnes, 1958, 1978). The classification system proposed by Varnes (1958, revised, 1978) is perhaps the most useful and widely adopted. An abbreviated version of Varnes classification system is presented in Table 2.4.

A comprehensive description of the various types of slope movement is outside the scope of this book. On the other hand, it is important to understand

TABLE 2.4   Abbreviated Classification of Slope Movement

		Type of Material	
		Engineering Soils	
Type of Movement	Bedrock	Predominantly Coarse	Predominantly Loose
Falls	Rock fall	Debris fall	Earth fall
Topples	Rock topple	—	—
Slides			
Rotational (*few units*)	Rock slump	Debris slump	Earth slump
Translational (*many units*)	Rock block slide	Debris block slide	Earth block slide
	Rock slide	Debris slide	Earth slide
Lateral spreads	Rock spread	Debris spread	Earth spread
Flows (creep)	Rock flow	Debris flow	Earth flow

Source: Adapted from Varnes (1978).

and to be able to identify various types of slope movement, not only for purposes of avoiding unstable slopes but also for designing prevention and control systems. Methods useful for prevention and control of one types of slope failure (e.g., shallow, translational sliding) may be ineffectual against deep-seated, rotational failures).

2.8.2 Causes of Slope Failure

The mass stability of slopes is governed by topographic, geologic, and climatic variables that control shear stress and shear resistance in a slope. Slopes fail when shear stress exceeds shear strength along a critical sliding surface. The factor of safety of a slope is defined as the ratio of shear strength to shear stress along a critical failure surface. The surface with the lowest ratio is the critical or failure surface. This surface also demarcates the boundary between stable and moving ground. The term "slide" implicitly specifies relative motion between the two. Methods of analyzing and computing the factor of safety of slopes against sliding are reviewed briefly in Section 2.9.

Any variable or factor that increases shear stress or conversely that decreases shear strength will tend to cause slope movement. Causes of slope instability or failure have been grouped into these two categories by Varnes (1958) as illustrated in Table 2.5. Timely identification and recognition of these factors is the key to prevention and control of slope movements.

Removal of lateral support by either natural or human agencies is probably the most common of all factors leading to instability. This observation suggests the use of buttress or retaining structures at the foot of slopes as a remedial measure. The addition of water to a slope may contribute simultaneously to an increase in stress and decrease in strength. Water has been implicated as either the primary or a major controlling factor in 95 percent of all landslides (Chassie and Goughnour, 1976).

Accordingly, drainage and diversion measures are without doubt among the mot effective means of preventing and/or controlling slope failures. Land-

---

TABLE 2.5   Causes of Slope Failure

Increase in Shear Stress
1. Surcharging slope (structures and fills at top).
2. Removal of lateral support (cuts and excavations at toe).
3. Rapid changes in water level adjacent to slope ("sudden drawdown").
4. Increase in lateral stress (water-filled cracks and fissures).
5. Earthquake loading (increases in horizontal or downslope driving forces).

Decrease in Shear Strength
1. Increased pore water pressure (PWP), which reduces effective stress (storm-water infiltration into slope, uncontrolled discharge of water from drains, earthquake-induced PWP).
2. Presence of swelling clays (uptake of water with loss of intrinsic cohesion).
3. Weathering and physicochemical degradation (ion exchange, hydrolysis, solutioning, etc.).
4. Progressive failure by shear strain softening.

slide remedial measures and their effectiveness have been discussed by Zaruba and Mencl (1969), Hutchinson (1978), and Schuster (1992).

Vegetation can ameliorate many of the factors and conditions that cause instability. Woody vegetation growing on a slope can reduce moisture stress via evapotranspiration, and increase overall stability as a result of buttressing and soil arching action. The contribution of slope vegetation to the mass stability of slopes is evaluated in detail in Chapter 3.

2.8.3   Indicators of Slope Instability

There are several visual indicators of hillside instability that are very useful for identifying areas of potential slope movement. Identification of unstable slopes (or slopes with a high landslide potential) can be made by using certain topographic, vegetative, hydrologic, and geologic indicators as summarized in Table 2.6. Most of these signs of hillside instability can be spotted by visual reconnaissance in the field. They are signals for precautions or preventative measures that should be employed during and after slope disturbance in the area.

Peck (1967) stated a postulate that is useful for identifying potentially unstable slopes; namely, if there is no evidence of old landslides in an area, then it is unlikely that moderate construction activity will start a new one. On the other hand, if old landslides abound, then it is quite likely that even minor construction operations will activate an old slide or cause a new one.

2.9   SLOPE STABILITY PREDICTIONS

2.9.1   Approaches to Analysis

Numerous methods are available for quantitative prediction of the stability of slopes and embankments. Two basic approaches are (1) limit-equilibrium analysis and (2) deformation analysis. Most methods in use today fall into the former category. Limit-equilibrium methods explicitly take into account the major factors that influence the shear stress and shear resistance of a slope (see Table 2.5). In addition, these methods are simpler to apply than deformation analysis. Limit-equilibrium methods do not result in calculation of expected slope deformations; however, there are many instances in which precise deformations are of minor concern.

A detailed treatment of slope stability analysis lies outside the scope of this book. The interested reader is referred to geotechnical texts on this subject (see Abramson et al., 2002; and Huang, 1983). The intent here is mainly to review basic concepts of stability analysis and to describe a few methods that are employed in practice. This review will (1) explain the combination of conditions that can result in slope failure; (2) identify the relative importance to stability of different soil, slope, and hydrologic variables; and (3) establish

TABLE 2.6   Features Indicating Landslides or Areas with High Landslide Potential

Feature	Significance
1. Hummocky, dissected topography	Common feature in old and active progressive slides (slides with many individual components). Slide mass is prone to gullying.
2. Abrupt change in slope	May indicate either an old landslide area or change in the erosion characteristics of underlying material. Portion with low slope angle is generally weaker and often has higher water content.
3. Scarps and cracks	Definite indication of an active or recently active landslide. Age of scarp can be estimated by the amount of vegetation established upon it. Width of cracks may be monitored to estimate relative rates of movement.
4. Graben, or "stair-step," topography	Indication of progressive failure. Complex or nested series of rotational slides can also cause surface of slope to appear stepped or tiered.
5. Lobate slope forms	Indication of former earthflow or solifluction.
6. Hillside ponds	Local catchments or depressions formed as result of feature 4, Graben topography, act as infiltration source, which can exacerbate or accelerate slope failure.
7. Hillside seeps	Common in landslide masses and areas with high landslide potential. Can usually be identified by associated presence of phreatophyte vegetation (e.g., equisetum, cattails, alder, willows, etc.) in vicinity of seep.
8. Incongruent vegetation	Patches or areas of much younger or very different vegetation (e.g., alder thickets) may indicate recent landslides or unstable ground.
9. "Jackstrawed" trees	Leaning or canted trees on a slope are indicators of previous episodes of slope movement or soil creep.
10. Bedding planes or joints	Potential surface of sliding for translational slope movements.

an analytical or quantitative basis for assessing the influence of vegetation on mass stability.

2.9.2   Limit-Equilibrium Analysis

Limit-equilibrium analysis is used to determine the factor of safety for a given slope; it can also be used to determine the effect of varying one or more parameters on the stability of a slope. A number of methods and procedures based on limit-equilibrium principles have been developed for this purpose.

Regardless of the specific procedure, the following principles (Morgenstern and Sangrey, 1978) are common to all methods of limit-equilibrium analysis:

- *A failure surface or mechanism is postulated.* In the simplest case, idealized slopes are assumed to fail along planes or circular sliding surfaces as shown schematically in Figure 2.5. More complex failure surfaces can also be proposed and analyzed when slope conditions are not uniform.
- *The shearing resistance required to equilibrate the failure mass is calculated by means of statics.* The potential failure mass is assumed to be in a state of "limiting equilibrium" and the shear strength of the soil or rock in the failure mass is mobilized everywhere along the slip surface.
- *The calculated shearing resistance required for equilibrium is compared with the available shear strength.* This comparison is made in terms of

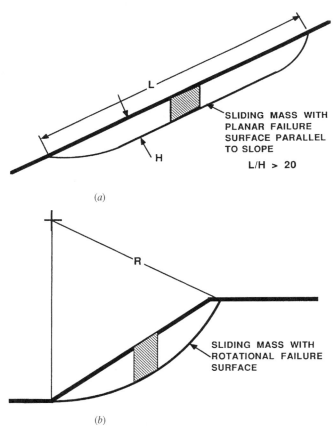

(a)

(b)

Figure 2.5   Slope-failure models or mechanisms for mass-stability analyses: (*a*) translational failure, "infinite slope" model, and (*b*) rotational failure, circular arc model.

the factor of safety, which is generally defined as the factor by which the shear strength parameter must be reduced in order to bring the slope into a state of limiting along a given slip surface.

- *The mechanism or slip surface with the lowest factor of safety is found by iteration.* The surface with the lowest computed safety factor is the critical slip surface. If the location of the slip surface is predetermined or constrained by stratigraphic control (e.g., a weak clay seam, relict-jointing surface, contact between residual soil and underlying bedrock, etc.), other trials are usually unnecessary.

The essentials for conducting a slope-stability analysis include the following requirements:

- Accurate description of slope geometry
- Reliable soil properties (c, ϕ)
- Correct definition of external loads, i.e., surcharge, line loads, earthquake loads
- Correct description of slope hydrology, i.e., phreatic surface (groundwater table, or GWT) and seepage conditions

These constraints and requirements are sometimes difficult to observe in practice. Some common methods of analysis do not rigorously meet all requirements for static equilibrium. This is not as serious a limitation, however, as errors introduced by uncertainty in the proper choice and selection of input parameters, particularly values for shear strength of soil (Singh, 1970). In spite of these drawbacks and problems, limit-equilibrium analysis provides powerful insights into the factors and conditions governing the stability of slopes; it also provides a rational basis for assessing slope hazard and designing remedial measures.

2.9.3 Shear-Strength Parameters

Determination of the factor of safety by limit-equilibrium methods requires an estimate of the shear resistance that can be mobilized along the assumed failure surface. The shear strength of soil or unconsolidated rock is given by the Coulomb failure criterion:

$$s = c + \sigma \tan \phi \tag{2.2}$$

where s = shear strength of material
$\quad \sigma$ = normal stress acting on the failure surface
$\quad \phi$ = angle of internal friction;
$\quad c$ = cohesion intercept.

The angle of internal friction (ϕ) and the cohesion (c) are known as the shear-strength parameters. They can be determined from various laboratory tests on representative samples of soil or, alternatively, back-calculated from analysis of a failed portion of a slope by assuming a factor of safety equal to unity.

2.9.4   Translational Slope Failures

A great number of slope failures can be characterized by shallow, translational sliding. Studies in both Britain and Japan have shown that 65 percent of all slope failures along road right-of-ways are essentially planar and occur at depths of one meter or less. These are also the types of slope failures most likely to be influenced by vegetation and by bio-stabilization treatments. Accordingly, we devote some attention to the analysis of translational failures. The reader is referred to other texts (see Huang, 1983, and Abramson, 2002) for analysis of deeper-seated, rotational failures. The stability of simple, natural slopes where all boundaries (ground surface, phreatic surface, and basal-sliding surface) are approximately parallel can be modeled and analyzed by so-called "infinite-slope" equations. In this analysis the slip surface is assumed to be a plane roughly parallel to the ground surface.

This type of analysis is appropriate when sliding takes place such that the ratio of depth to length of the sliding mass is small as noted in Figure 2.6.

The following types of slopes or slope conditions meet the aforementioned criteria:

1. Loose products of weathering (residual soil) overlying and inclined bedrock contact
2. Inclined planes of stratification dipping downslope that are underlain by stronger strata
3. Bedrock slopes mantled with glacial till or colluvium
4. Homogenous slopes of coarse-textured, cohesionless soil (e.g., sand dunes, sandy embankments, or fills)

In the first three slopes, the surface of sliding is predetermined by stratigraphic control. In the fourth, the slope failures are restricted to shallow, surface sloughing, because shear strength increases steadily with depth.

Because of the geometry of an infinite slope, overall stability can be determined by analyzing the stability of a single, vertical element in the slope as shown in Figure 2.6. End effects in the sliding mass can be neglected and so too can lateral forces on either side of the vertical element, which are assumed to be opposite and equal. The factor of safety based on an infinite-slope analysis for the conditions shown in Figure 2.6 is given by the following equation:

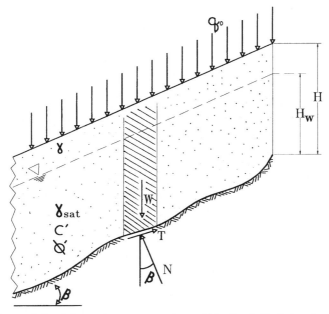

Figure 2.6   Schematic representation of idealized "infinite slope."

$$F = \frac{[c'/\cos^2 \beta \tan \phi' + (q_0 + \gamma H) + (\gamma_{\text{BUOY}} - \gamma)H_w]\tan \phi'/\tan \beta}{[(q + \gamma H) + (\gamma_{\text{SATD}} - \gamma)H_w]} \quad (2.3)$$

where ϕ' = effective angle of friction

$\quad\quad c'$ = effective cohesion intercept

$\quad\quad \beta$ = slope angle of natural ground

$\quad\quad \gamma$ = moist density of soil

$\quad\gamma_{\text{BUOY}}$ = buoyant density of soil

$\quad\gamma_{\text{SATD}}$ = saturated density of soil

$\quad\quad H$ = vertical thickness (or depth) of sliding surface

$\quad\quad H_w$ = piezometric height above sliding surface

$\quad\quad q_0$ = uniform vertical surcharge stress on slope

This expression is quite general and takes into account the influence of surcharge (q_0), the presence of a phreatic surface or groundwater table in the slope (H_w), and the existence of cohesion (c'). The influence of root reinforcement in the soil, which affects only the cohesion and not the soil friction, can be accounted for by adding a "root-cohesion" term (c_R) to the numerator of Equation (2.3) as explained in Chapter 3(Section 3.5.3). When the groundwater table or phreatic surface is parallel to the slope, as shown schematically in Figure 2.6, seepage also occurs parallel to the surface. An alternative expression can be derived to account for groundwater movement or seepage in other directions, for example, flowing vertically downward or emerging from the slope. These other directions may be important to consider as well, particularly in the case of slope vegetation and biostabilization measures, which can modify seepage direction near the face of a slope. The influence of seepage direction on the factor of safety is evaluated in Chapter 3.

A number of particular or special cases of interest can be derived from the general expression (Equation 2.3) for an infinite slope. These special cases include the following:

Case i: Cohesionless Slope, No Surcharge ($c' = 0$; $q_0 = 0$)

$$F = \frac{[\gamma(H - H_w) + \gamma_{\text{BUOY}} H_w] \tan \phi' / \tan \beta}{[\gamma(H - H_w) + \gamma_{\text{SATD}} H_w]} \qquad (2.4)$$

This equation shows the influence of a rise in the phreatic surface in a shallow, cohesionless soil overlying an inclined bedrock contact. Note that the controlling factor is not the absolute rise but rather the ratio (H_w/H). In other words, the same amount of rise in a phreatic surface is more serious in a "thin" soil mantle than a "thick" one.

Case ii: Saturated, Cohesionless Slope, No Surcharge ($c' = 0$; $q_0 = 0$; $H_w = H$)

$$F = \{\gamma_{\text{BUOY}} / \gamma_{\text{SATD}}\} \frac{\tan \phi'}{\tan \beta} \qquad (2.5)$$

But for most soils:

$$\gamma_{\text{BUOY}} \simeq 0.5 \gamma_{\text{SATD}}$$

Therefore:

$$F = 0.5 \frac{\tan \phi'}{\tan \beta}$$

This equation yields the factor of safety for the worst case of complete saturation of a cohesionless infinite slope. The factor of safety is approximately one-half that of the dry case.

Case iii: Dry, Cohesionless Slope, No Surcharge ($c' = 0$; $q_0 = 0$; $H_w = 0$)

$$F = \frac{\tan \phi'}{\tan \beta} \qquad (2.6)$$

This equation shows that in a dry, cohesionless material, the critical slope angle is equal to the angle of internal friction of the soil. If the material is end-dumped or side cast in a loose condition, this angle will be equivalent to the angle of repose.

Case iv: Stable Slopes with Cohesion, No Surcharge ($q_0 = 0$; $F \geq 1$)

(a) Dry Slope ($H_w = 0$):

$$c_d/\gamma H = F \cos^2 \beta\{\tan \beta - \tan \phi'\} \qquad (2.7)$$

where F = desired factor of safety; and
   c_d = required cohesion

(b) Saturated Slope ($H_w = H$):

$$c_d/\gamma H = F \cos^2 \beta [\tan \beta - (\gamma_{\text{BUOY}}/\gamma_{\text{SATD}}) \tan \phi'] \qquad (2.8)$$

These equations are useful for determining the amount of cohesion that must be present or developed—e.g., from root reinforcement, to achieve a desired factor of safety ($F \geq 1$) for a given depth of sliding (H), slope angle (β), and friction angle (ϕ'). Contributions to cohesion from root reinforcement and methods for estimating root cohesion are described in Chapter 3. Note that the required cohesion is directly proportional to the thickness of the sliding mass.

2.10 CONTROL OF MASS WASTING

Measures to control or prevent mass wasting fall in two basic categories, namely, measures that (a) decrease shear forces and (b) increase shear resistance. These two approaches are antidotes to the factors that cause mass slope failures (see Table 2.5). Schuster (1992) has presented a comprehensive review of slope stabilization approaches and techniques. This topic is also treated in considerable detail by Abramson et al. (2002).

Water is involved in the great majority of slope failures, because the presence of water both increases shear stresses and decreases shear strength. Accordingly, water drainage and diversion measures are extremely important in mass-slope stabilization.

Soil bioengineering measures (Gray and Sotir, 1996) increase stability mainly by increasing shear resistance either from reinforcement of the soil mantle by roots and/or from interception of shallow slip surfaces by imbedded stems. The contribution to increased shear resistance by woody roots is discussed further in Chapter 3. Some biostabilization measures also improve stability by modifying the hydrologic regime in the soil either by transpiring soil moisture or by acting as drains. Live fascines, elongated bundles of brush buried in shallow trenches, for example, can act as conduits to intercept and facilitate drainage (Gray and Sotir, 1996).

2.11   SLOPE-STABILITY AND CHANNEL-EROSION THRESHOLDS

2.11.1   Significance

Assessment of the spatial distribution of relative slope stability against mass movement and resistance to channel erosion or incision by overland flow is important not only in land-use decisions and planning in natural terrain but also in evaluating the effectiveness and permanence of artificial slope designs. The latter are particularly relevant with regard to proposed landform designs in: (1) mined land rehabilitation, (2) hillside housing developments, (3) transportation corridors in hilly terrain, (4) watershed restoration, and (5) waste-disposal landfills.

2.11.2   Approaches

Several different approaches to landslide hazard assessment of terrain have been developed, ranging from the empirical to the more mechanistic. These include: (1) classification into hazard categories based on field mapping of existing slope failures and professional judgment about the influence of local geology and topography; (2) classification based on correlation of terrain attributes (e.g., hillslope gradient, bedrock type, etc.) of areas mapped in the field with the incidence and character of slope failures; (3) use of slope stability theorems (e.g., infinite slope equation) applied to selected areas or individual sites on a hillside (see Section 2.9.2); and (4) coupling of digital representation of topography, a slope-stability model, and a subsurface-flow model to predict the spatial distribution of relative slope stability. The benefits and limitations of these approaches have been reviewed by Montgomery and Dietrich (1994). The advantages of the fourth approach [which includes a shallow landslide model known as Shallow Slope Stability Failure Analysis (SHALSTAB)] are identified and discussed by Dietrich et al., 2001

2.11.3   Slope-Stability Threshold

General Features: The spatial distribution of shallow landslides in a watershed or catchment can be predicted by a model known as SHAL-STAB. which couples a digital representation of the topography with a shallow, subsurface flow model and an infinite-slope stability model (Dietrich et al., 1992, 1993, 2001). A digital terrain model, TOPOG (O'Loughlin, 1986, Moore et al., 1988), is adapted for this purpose. TOPOG can be used to analyze landscape morphology in a manner that is strongly tied to its hydrologic response.

SHALSTAB is a relatively simple model that recognizes the importance of topography as expressed via topographic attributes that can be determined fairly easily. Past studies of shallow landslides have shown (Dietrich and Dunne, 1978; Reneau and Dietrich, 1987) that shallow landslides occur on steep slopes and often in areas of strong planform convergence where shallow subsurface flow may be concentrated. This finding suggests that surface topography is a primary indicator of where shallow slope failures are most likely to occur. The central importance of topography makes SHALSTAB of particular interest to landform grading and as a potential check on grading designs. Use of the model does not rely upon the input of often unreliable and difficult to measure soil properties

Digital Terrain Model: TOPOG assumes a steady-state rainfall, and it maps the spatial pattern of equilibrium soil saturation based on analysis of upslope contributing areas, soil transmissivity, and local gradient, as shown schematically in Figure 2.11. TOPOG draws contours and lines normal to the contours that extend up from low elevations to ridges and hill tops. The latter approximate flow lines for shallow subsurface and overland flow. The combination of two adjacent flow lines, intersecting two successive contours. divides the land surface into distinctly shaped elements, as shown in Figure 2.7.

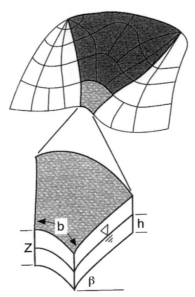

Figure 2.7 Topographic elements used in the program TOPOG are defined by the intersections of contours and flow-line boundaries. The upslope, contributing area (shaded) is the cumulative drainage area of all topographic upslope elements draining into the element (from Montgomery and Dietrich, 1994).

The resulting map or diagram resembles a flow net constructed atop a piezometric surface in a watershed. For each element, the total contributing area, a, can be calculated as well as the ratio a/b from the bottom contour length of the element.

A conventional topographic map of a 1.2-square-kilometer watershed in Northern California (Tennessee Valley) is shown in Figure 2.8. This map also shows the location and distribution of landslides, channels, and interstratified alluvial and debris flow deposits in major valleys. The corresponding derivative map generated by TOPOG is shown in Figure 2.9. The contour lines on this derivative map closely track those on the corresponding topographic map (Dietrich et al. 1993). Based on the element shape created by TOPOG, each element has been classified

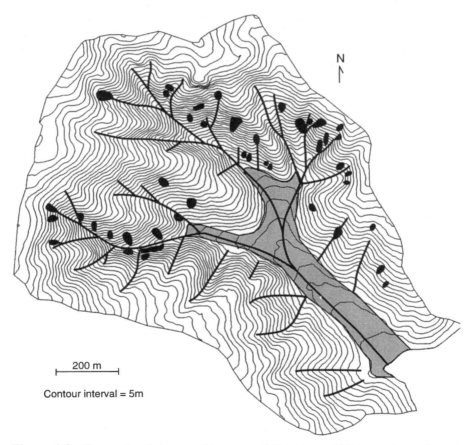

Figure 2.8 Conventional topographic map of Tennessee Valley study catchment showing distribution of landslides (black areas), channels (black lines), and interstratified alluvial and debris flow deposits in major valleys (shaded area) (from Montgomery and Dietrich, 1994).

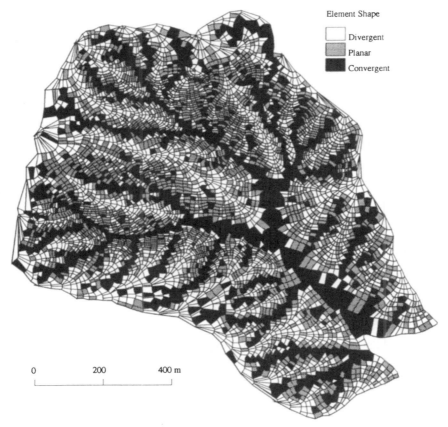

Figure 2.9   TOPOG map of Tennessee Valley study catchment showing spatial pattern of convergent, planar, and divergent topographic elements (from Dietrich et.al., 1993).

(Dietrich et al., 1993) into divergent, planar, and convergent morphology as shown in Figure 2.9. This classification successfully identifies the hillside swales, channels, and valley axes as convergent and the ridge-lines as divergent.

Shallow Subsurface Flow: Subsurface flow is assumed to occur parallel to the ground. Runoff that cannot be transmitted in the subsurface region must travel overland, and does so with a mean velocity u for depth, d, across the contour length, b, as shown in Figure 2.10. The following expression can be derived (Dietrich *et. al.*, 1993) based on the conservation of mass for steady-state, shallow subsurface flow and saturation overland flow (SOF):

$$qa - TMb = udb \qquad (2.9)$$

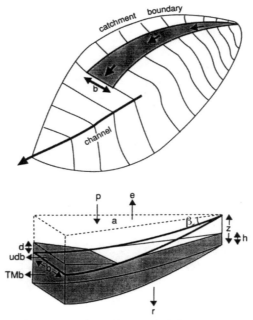

Figure 2.10   Plan and cross section of area draining across contour length, b, as defined by "flow lines" generated by the program TOPOG. The heavy line in the cross section depicts the ground surface. The stippled area is the shallow subsurface flow and saturation overland flow (SOR) represented by TMb and udb, respectively in Equation (2) (from Dietrich et.al., 1993).

where a = catchment draining across a contour of length, b;
$\quad\quad q$ = total runoff per unit area (viz., precipitation, p, minus the evaporation, e, minus deep drainage, r);
$\quad\quad T$ = transmissivity;
$\quad\quad M$ = slope gradient (equal to sin β),
$\quad\quad \beta$ = local slope angle;
$\quad\quad u$ = mean overland flow velocity;
$\quad\quad d$ = flow depth.

Some explanation is required for the use of sin β in lieu of the normal tan β for denoting the slope gradient. The gradient-driving flow in porous media is actually the hydraulic gradient. The hydraulic gradient in a slope is the difference in the piezometric head elevation divided by the slope distance over which the piezometric head elevation difference (head drop) occurs. This ratio is expressed by sin β. For flow parallel to the ground surface, the slope of the piezometric surface and ground surface are identical; hence M = sin β.

Equation 2.9 states that at steady state the difference between the total runoff and the saturated, shallow subsurface flow is equal to the overland flow. According to this equation, the ground will saturate everywhere the instant that overland flow begins. The point of saturation is defined, therefore, by the following:

$$udb \geq 0 \qquad (2.10)$$

Substituting Equation (2.9) into (2.10) yields the following expression:

$$a/b \geq (T/q)\, M \qquad (2.11)$$

Equation (2.11) can be rearranged to yield the following expression:

$$(a/bM) \geq (T/q) \qquad (2.12)$$

The equation states in effect that when the *topographic ratio, a/bM*, is greater than or equal to the "hydrologic ratio," T/q, the ground will saturate. The significance of this expression can be explained (Dietrich et al., 1993) as follows:

- a/bM is the topographic term that represents the tendency for flow convergence, i.e., [high a/b] and for the flow to travel quickly as shallow subsurface flow [high M]. Large areas draining to gentle slopes, i.e., [high a/b and low M] are prone to saturation.
- T/q is the hydrologic term that can be viewed as the ability of the subsurface to transmit water flow relative to the applied runoff. When this ratio is small—e.g., large, q, and small, T—the ground tends to saturate.

Slope Stability Threshold: As discussed previously (see Section 2.8.2), earthen slopes fail when the shear stress (due to the downslope weight of soil) exceeds the shear strength or resistance along a critical sliding surface. Mass stability is strongly dependent on pore pressures in the soil; failure occurs when pore pressure increases sufficiently along a critical sliding surface to reduce the shear resistance below the developed shear stress.

Shallow slope failures are the most common type of slope failure (see Section 2.9.4). They typically occur in sandy soils (which are weakest near the surface) and/or where some type of near-surface control limits the depth of the failure surface. A shallow underlying bedrock interface or other physical discontinuity will frequently provide such a limit. The underlying bedrock (or other discontinuity) is normally stronger and generally has a lower hydraulic conductivity than the overlying soil. The "infinite"-slope stability equation can be used to determine the factor of safety against failure under these conditions. The assumptions and characteristics of this equation have been described previously (see Section 2.9.4).

SHALSTAB assumes that the zone of shallow surface storm runoff is confined to a cohesionless soil overlying a less conductive but stronger bedrock. Accordingly, the dimension, H (see Figure 2.4), defines not only the maximum thickness of shallow surface flow but also the thickness of the shallow, potentially unstable surface layer. The dimension, H, defines the thickness of the subsurface, saturated flow zone. The surface layer does not have to be saturated for failure to occur, and in some cases (e.g., low slope gradients). this condition may not trigger failure. The transmissivity, T, of the surface soil layer is given by the following expression:

$$T = K \cdot H \cos \beta \tag{2.13}$$

where T = transmissivity;
  H = vertical thickness of shallow, potentially unstable soil layer ($H \cos \beta$ is the thickness measure normal to the slope);
  β = local slope angle of the ground surface

Substitution of Equation (2.13) into (2.9) yields an expression for the ratio of the actual runoff, qa, to the maximum subsurface flow, bTM, for the condition of *subsurface flows equal to or less than saturated* as follows:

$$\frac{qa}{bTM} = \frac{K \sin \beta \, H_w \cos \beta}{K \sin \beta \, H \cos \beta} = \frac{H_w}{H} \tag{2.14}$$

The thickness of the saturated zone, H_w, required to convey the imposed runoff varies from a small value up to H, thus the ratio, H_w/H, can vary from nearly zero to 1.0. Under fully saturated conditions—i.e., $H_w/H = 1.0$, Equation (2.14) reduces to the following:

$$a/b = (T/q) M = (T/q) \sin \beta \tag{2.15}$$

Equation (2.15) defines a boundary (or line) separating saturated and unsaturated catchment elements in a plot of a/b versus M. The location or position of this boundary in such a plot will depend on the value assigned to the hydrologic term (T/q).

For groundwater flow parallel to the slope, no surcharge, cohesionless soils ($c = 0$), and moist soil unit weight approximately the same as the saturated unit weight ($\gamma \approx \gamma_{sat}$) the infinite slope stability equation can be written (see Section 2.9.4) as:

$$F = \frac{(\gamma H - \gamma_w H_w)}{\gamma H} \frac{\tan \phi}{\tan \beta} \tag{2.16}$$

where F = factor of safety against failure;
γ = moist unit weight of soil;
γ_w = unit weight of water,
ϕ = angle of internal friction of soil

At failure, $F = 1$, and Equation (2.16) reduces to the following form:

$$\gamma H \tan \beta = (\gamma H - \gamma_w H_w) \tan \phi \qquad (2.17)$$

Dividing each side of Equation (2.10) by $\gamma H \tan \beta$ and rearranging terms yields:

$$1 = \{1 - (\gamma_w/\gamma) \, H_w/H\} \tan \phi/\tan \beta \qquad (2.18)$$

Substitution of Equation (2.18) into Equation (2.14) for H_w/H yields the following expression:

$$a/b \geq \gamma/\gamma_w \, (1 - \tan \phi/\tan \beta) \, (T/q) \, M \qquad (2.19a)$$

This equation states that instability will occur when the area per unit contour length, a/b, is greater than or equal to the product of four terms, namely, the unit weight ratio, γ/γ_w, the ratio of the tangents of the friction angle and ground surface angle, $\tan \phi/\tan \beta$, the hydrologic ratio, (T/q), and gradient of the ground surface, M (or $\sin \beta$). This expression can be simplified further by noting that for most soils the saturated density is approximately twice the density of water (viz., $\gamma_s \approx 2 \cdot \gamma_w$). Thus, instability can be expected when:

$$a/b \geq 2(1 - \tan \phi/\tan \beta) \, (T/q) \, M \qquad (2.19b)$$

For simplicity, the contribution of either intrinsic or apparent cohesion from root reinforcement is neglected. This makes the model conservative, i.e., it overpredicts areas of instability. A more sophisticated version of SHALSTAB has been derived by Montgomery and Dietrich (1994) that takes cohesion into account, but this requires additional information about soil depth or, at least, the depth of rooting. In addition, this relationship (Equation 2.19) is only valid for seepage flow parallel to the ground surface; it does not account for pore pressures associated with exfiltrating groundwater. This makes the model slightly less conservative, because exfiltrating flow leads to lower factors of safety.

This version of SHALSTAB emphasizes the importance of topography, because the model essentially holds all soil properties (thickness, unit weight, shear strength, and hydraulic conductivity) constant in space and then defines the topographic conditions that produce shallow slope failure for a given set of hydrologic conditions. As Dietrich et al. (1993)

note, the simplicity of this model is consistent with the normal lack of detailed knowledge about the spatial variability of soil properties in natural catchments.

Because of the parallel flow condition, slopes greater than or equal to the angle of internal friction of the soil are chronically (and unconditionally) unstable, regardless of level of saturation. Consideration of the infinite-slope equation can also establish circumstances when the slope will be stable even when fully saturated. These two limiting conditions are noted as follows:

- Unconditionally Stable:

$$\tan \beta \le [(\gamma H - \gamma_w H_w)/\gamma H] \tan \phi \qquad (2.20)$$

- Unconditionally Unstable:

$$\tan \beta \ge \tan \phi \qquad (2.21)$$

Equation (2.19) must be used to determine local stability for intermediate conditions. The topographic terms a/b and M capture the essential effects of topography on runoff. The ratio, a/b, reflects the topographic convergence that concentrates subsurface flow and boosts pore pressures, while M represents the influence of the ground slope. The steeper the ground, the faster the subsurface flow and the lower the relative wetness (or level of saturation) defined by H_w/H. The hydrologic ratio q/T reflects the magnitude of a precipitation event, as represented by q, relative to the subsurface ability to convey the water downslope, represented by transmissivity, T. The larger q relative to T, the more likely the ground is to saturate and, consequently, the greater the number of sites on a hillslope that will become unstable. In other words, the H_w/H specified by Equation (2.14) exceeds the H_w/H obtained by substituting Equation (2.14) into Equation (2.19), namely,

$$(H_w/H)_{crit} = (H_w/H) \ge \gamma/\gamma_w (1 - \tan \phi/\tan \beta) \qquad (2.22a)$$

If assume that $(\gamma \approx \gamma_w)$, then

$$(H_w/H)_{crit} = (H_w/H) \ge 2 (1 - \tan \phi/\tan \beta) \qquad (2.22b)$$

These two expressions give the proportion of the soil column—$(H_w/H)_{crit}$—that is saturated and in an unstable state, sometimes referred to as the degree of "wetness" (Montgomery and Dietrich, 1994). The saturation ratio or wetness cannot exceed 1.0 in the model. If $(H_w/H) \ge 1.0$ the excess water simply runs off as overland flow.

The stability categories defined by Equations (2.19), (2.20), and (2.21) can be shown on a plot of drainage area per unit contour length, a/b, versus slope gradient, tan β, as shown schematically in Figure 2.11. Note that, in this case, the slope gradient is represented in the conventional manner as the slope of the ground surface, tan β. Logarithmic axes are used to facilitate plotting the entire range of possible values. The threshold of saturation as defined by Equation (2.15) is shown as a diagonal line on the plot.

The four stability classes that describe the elements within a catchment or watershed for a particular hydrologic event are:

- UNCONDITIONALLY UNSTABLE: Elements predicted to be unstable even when dry (Equation 2.21).

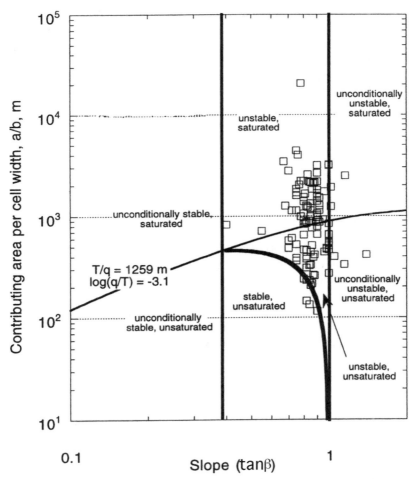

Figure 2.11 Delineation of slope-stability fields on a plot of contributing areas per unit contour length (a/b) versus slope of ground surface (tan β). Diagonal line represents the threshold of ground saturation for a particular value of the hydrologic parameter (q/T) (from Dietrich et al., 2001).

- UNSTABLE: Elements predicted to fail according to Equation 2.19, namely, $a/b \geq 2\,(1 - \tan\phi/\tan\beta)\,(T/q)\,M$.
- STABLE: Elements predicted NOT to fail according to Equation 2.19: $a/b < 2\,(1 - \tan\phi/\tan\beta)\,(T/q)\,M$. These are elements that have insufficient catchment area (and/or) degree of saturation to fail.
- UNCONDITIONALLY STABLE: Elements predicted to be stable even when fully saturated (Equation 2.20).

The stability of each topographic element in a watershed can thus be determined by plotting its position on a diagram similar to Figure 2.11 and noting its position relative to the boundaries discussed above. Assigning different values to the hydrologic-ratio term (T/q) allows simulations to determine the effect of storms of different intensities on stability and the spatial location and distribution of slope failures in a watershed. Conversely, it is possible to use the model to determine the critical rainfall necessary to cause instability for each topographic element.

Field Study Sites: The SHALSTAB predictions have been evaluated and tested in several watersheds in the western United States (Dietrich et al., 1993; Dietrich, 1999; and Montgomery and Dietrich, 1994) and have been found to yield reliable results and useful findings. The results of simulations in only one, well-studied watershed—the Tennessee Valley study site, located in Marin County in Northern California—are briefly reviewed herein. A topographic map of this catchment with the mapped distribution of landslides, channels, and channel deposits was shown earlier (see Figure 2.8). The corresponding map generated by the program TOPOG showing the spatial distribution of planar, convergent, and divergent elements in the catchment was also shown earlier (see Figure 2.9).

The use of Equation (2.19) to discriminate between STABLE and UNSTABLE areas requires a determination or assignment of values to the following hydrologic and/or soil parameters: (T/q), (γ/γ_w), and $\tan\phi$. Previous studies suggest that the coarse-textured soils at the study site have angles of internal friction in the range of 40 to 45 degrees and are generally cohesionless. Roots may contribute some apparent cohesion, but this influence can only be accounted for, partially, by using a high-end estimate for the friction angle. For plotting purposes, a range of 35 to 45 degrees could be adopted to delineate the range of likely thresholds of stability. Based on a number of other studies, Montgomery and Dietrich (1994) report the following most likely values for the hydrologic and soil variables in this watershed: transmissivity, $T = 17$ sq m/day, soil friction angle, $= 40$ degrees, and saturated soil density, $\gamma_{sat} = 18$ kN/cu m. These values are assumed to reasonably represent the natural heterogeneity within the catchment.

A plot of a/b vs. tan β for each mapped element in the study site catchment is shown in Figure 2.12. The elements are differentiated with regard to morphological shape (see Figure 2.9) as to whether they are convergent or divergent. Three different thresholds of saturation ($T/q =$ 175, 350, and 3500 m) and one stability-threshold limit ($\beta = 40$ degrees) are superimposed on the plot in Figure 2.12. The stability boundaries (between STABLE and UNSTABLE) and threshold-saturation limit shifts as this ratio is varied. At higher and higher rainfall rates more and more topographic elements become unstable.

Convergent topographic elements are plotted as circles, divergent elements as crosses in Figure 2.12. Overland flow rarely if ever occurs on topographically divergent hillsides (hilltops, ridges, and interfluve areas). Consequently, a reasonable upper limit to steady-state rainfall for use in slope-stability simulations is that which exceeds the amount required to saturate a slope and cause saturation to expand into divergent portions of the landscape. For the Tennessee Valley catchment, this rainfall is about 50 mm/d ($T/q = 350$ m). The predicted pattern of ground saturation for a T/q of 350 m is shown in Figure 2.13. Notice that this predicted pattern of ground saturation coincides very closely with the spatial pattern of convergent topographic elements in the TOPOG map of the Tennessee Valley catchment (see Figure 2.9). The critical or upper rainfall limit can also be used as a proxy for the steady-state rainfall required to predict zones of potential instability in this watershed. The

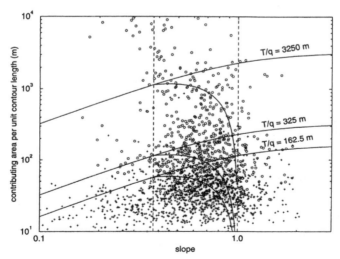

Figure 2.12 Plots of contributing area per unit contour length (a/b) vs. slope (tan β) for convergent (circles) and divergent (crosses) elements in the Tennessee Valley study site (after Montgomery and Dietrich, 1994).

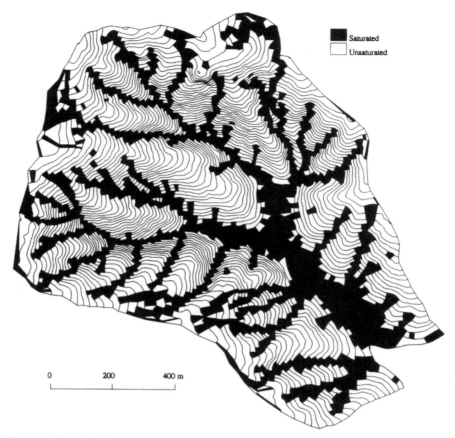

Figure 2.13 Predicted pattern of ground saturation (black areas) for a T/q value of 350 m during steady-state runoff in the Tennessee Valley study site (from Dietrich et.al., 1993).

majority of observed landslides in this catchment lie within zones of predicted instability determined in this fashion as shown in Figure 2.14.

2.11.4   Threshold of Erosion by Saturation Overland Flow

Referring back to the topographic element shown in Figure 2.10, at the instant of saturation, some excess overland flow can occur across the top of the element. This excess flow with a depth (d) is referred to as saturation overland flow (SOF). This overland flow tends to erode the element and initiate a channel by exerting a boundary shear or tractive stress (τ_b) on the element top surface. Following the analysis of Dietrich et al. (1993), channel incision is assumed to occur once a critical boundary shear stress (τ_c) is exceeded.

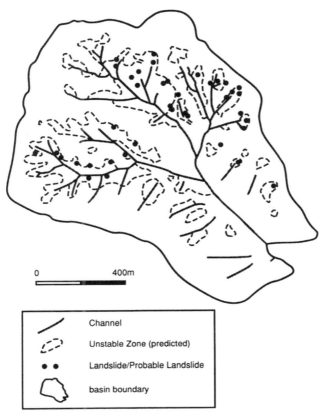

Figure 2.14   Predicted and observed pattern of shallow landslides for a T/q of 350 m, and angle of internal friction of 40 degrees (from Dietrich et al., 1993).

Sediment transport (q_s) is also assumed to be proportional to an excess boundary shear stress, according to the following functional relationship:

$$[q_s \propto (\tau_b - \tau_c)^n]$$

Thus, channels would be expected to form when ($\tau_b \geq \tau_c$) and tend to thin and disappear if $[\partial\,(\tau_b - \tau_c)^n/\partial x] \geq 0$. The limitations of this formulation notwithstanding, the tendency for scour (or deposition) should still respond to spatial change in excess of shear stress.

The boundary shear (or tractive) stress for steady flow is given by the following expression (Gray and Sotir, 1996):

$$\tau_b = \gamma_w \cdot d \cdot M \tag{2.23}$$

The density of water (γ_w) is known and the local slope (M) can be determined from a digital elevation model, as explained previously (see Section

2.11.3). The flow depth (d) must be calculated from the runoff rate and surface roughness. Equation (2.10), which apportions the total runoff between shallow subsurface flow and saturation overland flow, can be rewritten to solve for flow depth as follows:

$$d = \frac{1}{ub} \cdot [qa - TMb]$$

multiplying each side by $\gamma_w \cdot M$ and substituting into Equation (2.23) yields:

$$\tau_b = \gamma_w \cdot d \cdot M = \frac{\gamma_w M}{ub} \cdot [qa - TMb] \tag{2.24}$$

The velocity term (u) can be eliminated from Equation (2.24) by substituting an expression for overland flow (Darcy-Weisbach equation). Making this substitution, setting the boundary stress (τ_b) equal to the critical shear stress (τ_c), and rearranging terms yields the following equations (Dietrich et al., 1993) for laminar and turbulent overland flow respectively:

Laminar flow:

$$a/b \geq \frac{1.4g^{1/2}}{(\gamma_w)^{3/2}} \frac{(\tau_c)^{3/2}}{\nu M f^2 q} + (T/q) M \tag{2.25a}$$

Turbulent flow:

$$a/b \geq \frac{2g}{(\gamma_w)^3} \frac{(\tau_c)^3}{K\nu M^2 q} + (T/q) M \tag{2.25b}$$

where f = friction factor
ν = kinematic viscosity
g = acceleration constant of gravity
K = constant for given roughness condition

The above analysis indicates that channel initiation is controlled strongly by the critical shear stress (τ_c). Once the critical shear stress is exceeded, then the area per unit contour length (a/b) that is upslope of the channel varies directly with the critical shear stress raised to some power and inversely with roughness (K), precipitation (q), and if critical shear stress is sufficiently large, inversely with the slope (M).

If the critical shear stress is very small, then the first term in Equation 2.25 becomes insignificant and Equation 2.25 reduces to Equation (2.11), namely,

$$\text{If } \frac{2g(\tau_c)^3}{(\gamma_w)^3 K\nu M^2 q} << (T/q) M \text{ then } a/b \geq (T/q) M$$

This implies that if the critical shear stress is very low, channel initiation is equivalent to (or occurs upon the onset) of ground saturation. Dietrich et al. (1993) suggest that overland flow, draining a large area, can initiate a channel but once incision has occurred, seepage erosion (piping) can then advance the channel head upslope to the point where the ground is no longer saturated. Montgomery and Dietrich (1989) aptly note that "channel head locations are controlled by the competing influences of fluvial transport in channels and diffusional hillslope transport processes which concentrate colluvium in hollows and tend to infill channels."

2.11.5   Stability Fields and Threshold Boundaries

Several efforts have been made to establish erosion thresholds and to delineate boundaries for both slope failures (landsliding) and channel initiation by surface overland flow (SOF) in a given catchment or watershed. Dietrich et al. (1992) assigned reasonable values to the hydrologic and soil and slope variables in Equations (2.19) and (2.25) for the experimental Tennessee Valley watershed (see Figure 2.8). The stability fields and erosion thresholds for the dominant hydrologic and erosion processes are shown in Figure 2.15. Graphs

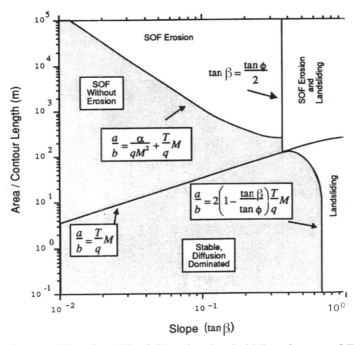

Figure 2.15   Definition of stability fields using threshold lines for case of $T/q = 350$ m, $\phi = 35°$, $K = 10,000$, $\tau_c = 160$ dynes/sq cm, and $q = 0.05$ m/d. Shaded areas are stable. Diagonal line represents boundary between saturated and unsaturated topographic elements (adapted from Dietrich et al., 1993.

like this provide a different way to examine a landscape and to define quantitatively the spatial dominance of different erosion processes in a watershed.

All the topographic elements (planar, divergent, and convergent) in a given catchment or watershed (see Figure 2.9) can be plotted on a graph similar to the one shown in Figure 2.15. This plot will determine whether an element lies in a stable or unstable field and what type of erosion process (or combination of processes) acts on the element. Figure 2.16 shows the spatial distribution and significance of different erosional processes determined in this manner. The ridgelines are dominated by slope-dependent sediment processes that feed sediment to landslide prone planar and subtle, convergent slopes bordering the ridges and valley swales. Surface runoff in the larger valleys then conveys the sediment in channels further downstream.

2.12 SUMMARY

Superficial erosion and mass wasting shape much of the topography of the Earth's surface. Surficial erosion entails the detachment and transport of in-

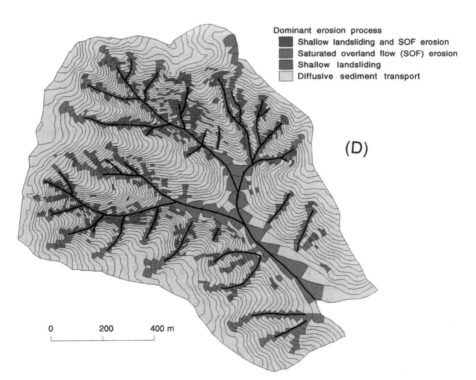

Figure 2.16 Predicted spatial pattern of dominant erosion processes in Tennessee Valley experimental watershed with $\tau_c = 160$ dynes/sq cm (from Dietrich et al., 1993).

dividual particles; mass wasting, the movement of large, intact masses of soil and/or rock along critical failure surfaces. Many of the same slope, soil, and hydrologic factors that control surficial erosion also control mass wasting (e.g., steepness of slope and shear strength of soil). The two processes differ, however, in several important respects.

Rainfall erosion is controlled by four basic factors, namely, climate, soil type, topography, and vegetative cover. The susceptibility of a soil to erosion is known as its "erodibility." Some soils (e.g., silts and silty sands) are inherently more erodible than others (e.g., coarse, well-graded gravels).

Surficial erosion can be controlled or prevented by observing some basic principles that are enumerated herein. These principles are universally applicable; they should be observed regardless of whether conventional or soil bioengineering treatments are contemplated.

The mass stability of slopes is governed by topographic, geologic, and climatic variables that control shear stress and shear resistance in a slope. Slopes fail when shear stress exceeds shear strength along a critical sliding surface. Unstable slopes or slopes with a high landslide potential can be identified using certain topographic, vegetative, hydrologic, and geologic indicators as summarized in Table 2.6.

A great number of slopes fail by shallow, translational sliding. Studies in both Britain and Japan have shown that 65 percent of all slope failures along road right-of-ways are essentially planar and occur at depths of one meter or less. These types of slope failures are those most likely to be influenced by vegetation and by soil bioengineering treatment. The stability of simple, natural slopes where all boundaries (ground surface, phreatic surface, and basal sliding surface) are approximately parallel can be modeled as "infinite" slopes. In this analysis, the slip surface is assumed to be planar and roughly parallel to the ground surface.

Measures to control or prevent mass wasting fall in two basic categories, namely, measures that (a) decrease shear forces and (b) increase shear resistance. Water is involved in the great majority of slope failures, because the presence of water both increases shear stresses and decreases shear strength. Accordingly, water drainage and diversion measures are extremely important in mass slope stabilization. Soil bioengineering measures increase stability mainly by increasing shear resistance either from reinforcement of the soil mantle by roots and/or from interception of shallow slip surfaces by imbedded stems. Some biostabilization measures also improve stability by modifying the hydrologic regime in the soil either by transpiring soil moisture or by acting as drains.

Assessment of the spatial distribution of relative slope stability against mass movement and resistance to channel erosion or incision by overland flow is important not only in land-use decisions and planning in natural terrain but also in evaluating the effectiveness and permanence of different man-made slope designs. The spatial distribution of shallow landslides in a watershed can be predicted by a model known as SHALSTAB, which couples a digital representation of the topography with a shallow subsurface flow model and an infinite-slope stability model. SHALSTAB is a relatively simple model that recognizes the importance of topography as expressed via topographic attributes that can be determined fairly easily.

At the instant of saturation, some excess overland flow can occur across the top of a topographic element in a watershed. This overland flow tends to erode the element and initiate a channel by exerting a boundary shear or tractive stress (τ_b) on the element top surface. Following the analysis of Dietrich et al. (1993), channel incision is assumed to occur once a critical boundary shear stress (τ_b) is exceeded.

Erosion thresholds can be established and boundaries delineated for both slope failures (landsliding) and channel initiation by surface overland flow in a given catchment or watershed. All the topographic elements (planar, divergent, and convergent) in a given watershed can be plotted on a graph. This plot will determine whether an element lies in a stable or unstable field and what type of erosion process (or combination of processes) acts on the element. Graphs like this provide a different way to examine a landscape and to define quantitatively the spatial dominance of different erosion processes in a watershed.

2.13   REFERENCES

Abramson, L. W., T. S. Lee, S. Sharma, S., and G. M. Boyce. 2002. *Slope Stability and Stabilization Methods.* 2nd ed. New York: John Wiley & Sons.

Bagnold, R. A. 1941. *The Physics of Wind Blown Sand and Desert Dunes.* London: Methuen.

Beer, C. E., and H. P. Johnson. 1963. Factors in gully growth in deep loess area of Western Iowa. *Transactions ASAE* 6 (3):237–240.

Chassie, R. G., and R. D. Goughnor. 1976. National highway landslide experience. *Highway Focus* 8 (1):1–9.

Chepil, W. S., and N. P. Woodruff. 1963. The physics of wind erosion and its control. *Advances in Agronomy* 15: 211–302.

Dietrich, W. E. and T. Dunne. 1978. Sediment budget for a small catchment in mountainous terrain. *Zeitschrift für Geomorphologie, Supplemental* 29:191–206.

Dietrich, W. E., C. J. Wilson, D. R. Montgomery, J. MacKean, and R. Bauer. 1992. Erosion thresholds and land surface morphology. *Geology* 20: 675–679.

Dietrich, W. E., C. J. Wilson, D. R. Montgomery, and J. MacKean. 1993. Analysis of erosion thresholds, channel networks and landscape evolution using a digital terrain model. *Journal of Geology* 101:259–278.

Dietrich, W. E., D. Bellugi, R. Real de Asua. 2001. Validation of the shallow landslide model, SHALSTAB, for forest management. In *Land Use and Watersheds: Human Influence on Hydrology and Geomorphology in Urban and Forest Areas, eds.* M. S. Wignosta and S. J. Burges, 195–227. *Water Science and Applications 2. American Geophysical Union.*

Ellison, W. D. 1948. Erosion by raindrop. *Scientific American* August: 1–7.

Emerson, W. W. 1967. A classification of soil aggregates based on their coherence in water. *Australian Journal of Soil Research* 2:211–217.

Goldman, S. J., K. Jackson, and T. A. Bursztynsky. 1986. *Erosion and Sediment Control Handbook*. New York: McGraw-Hill.

Gray, D. H., and R. Sotir. 1996. *Biotechnical and Soil Bioengineering Slope Stabilization*. New York: John Wiley & Sons.

Huang, R. 1983. *Stability of Earth Slopes*. New York: Van Nostrand Reinhold.

Hutchinson, J. N. 1978. Assessment of the effectiveness of corrective measures in relation to geologic conditions and types of slope movement. *Bulletin International Association of Engineering Geologists* 16:131–155.

Israelson, E. 1980. Erosion control during highway construction—Manual on principles and practices. *Transportation Research Board Program Report No. 221*. Washington, DC: National Academy of Sciences (NAS) - National Research Council (NRC), Transportation Research Board.

Kay, B. L. 1983. Straw as an erosion control mulch. *Agronomy Progress Report No. 140*. Davis, CA: Univ. of California Davis Agricultural Experiment Station.

Keown, M. P. et al. 1977. Literature survey and preliminary evaluation of streambank protection methods. *Technical Report* H-77-9. Vicksburg, MS: U.S. Army Waterways Experiment Station.

Leopold, L. B. 1994. *A View of the River*. Cambridge, MA: Harvard University Press.

Montgomery, D. R., and W. E. Dietrich. 1994. A physically based model for the topographic control on shallow landsliding. *Water Resources Research* 30 (4):1153–1171.

Moore, I. D., E. M. O'Loughlin, and G. J. Burch. 1988. A contour-based topographic model for hydrological and ecological studies. *Earth Surface Processes and Landforms* 13:305–320.

Morgenstern, N. R. and D. A. Sangrey. 1978. Methods of stability analysis. In "Landslides: Analysis and Control," ed. R. L. Schuster and R. J. Krizek, 155–171, *Transportation Research Board Special Report 176*. Washington, DC: NAS-NRC.

National Cooperative Highway Research Program (NCHRP). 2005. Environmentally Sensitive Channel and Streambank Protection Measures, NCHRP Report 544. Washington, DC: Guidance manual prepared for the NCHRP.by John McCullah and Donald Gray, Salix Applied Earthcare, Redding, CA.

O'Loughlin, E. M. 1986. Prediction of surface saturation zones in natural catchments by topographic analysis. *Water Resources Research* 22:794–804.

Peck, R. 1967. Stability of natural slopes. *Journal of Geotechnical Engineering (ASCE)* 93 (4):437–451.

Renard, K. G., G. R. Foster, G. A. Weesies, and J. P. Porter. 1991. RUSLE: Revised universal soil loss equation. *Journal of Soil and Water Conservation* 46:30–33.

Reneau, S. L., and W. E. Dietrich. 1987. The importance of hollows in debris flow studies; examples from Marin County, California. *Geological Society of America Reviews in Engineering Geology* 7:165–180.

Schuster, R. L. 1992. Recent advances in slope stabilization. Keynote paper, Session G.3 (Stabilization and Remedial Works). *Proceedings,* 6th Intl. Conference on Landslides, Christchurch, New Zealand, Feb. 10–14.

Schwab, G. O. et al. 1966. *Soil and Water Conservation Engineering,* New York: John Wiley & Sons.

Sherard, J. L. et al. 1978. Pinhole test for identifying dispersive soils. *Journal of Geotechnical Engineering (ASCE)* 102 (GT1):69–85.

Singh, A. 1979. Shear strength and stability of man-made slopes. *Journal. of Soil Mechanics and Foundations Division (ASCE)* 96 (SM6):1879–1890.

USDA Soil Conservation Service. 1972. *Procedures for Computing Sheet and Rill Erosion on Project Areas.* Technical Release No. 51. Washington, DC: U.S. Department of Agriculture.

USDA Soil Conservation Service. 1978. *Predicting Rainfall Erosion Losses; A guide to conservation planning.* USDA Handbook No. 537. Washington, DC: U.S. Department of Agriculture.

U.S. Department of Navy. 1971. *Design Manual DM-7 for Soil Mechanics, Foundations and Earth Structures.* Washington, DC: Naval Facilities Engineering Command.

Varnes, D. J. 1958. Landslide types and processes. In "Landslides and Engineering Practice," ed. E. B. Eckel, 20-47, *Highway Research Board Special Report No. 29.* Washington, DC: NAS-NRC.

Varnes, D. J. 1978. Slope movements, types and processes. In "Landslides, Analysis and Control," ed. R. L. Schuster and R. J. Krizek, 11–13, *Transportation Research Board Special Report. No. 176.* Washington, DC: NAS-NRC.

Volk, G. M. 1937. Method of determining the degree of dispersion of the clay fraction of soils. *Proceedings Soil Science Society of America* 2:432–445.

Wischmeier, W. H., C. B. Johnson, and B. V. Cross (1971). A soil erodibility nomograph for farmland and construction sites. *Journal of Soil and Water Conservation* 26(5): 189-193.

Wischmeier, W.H. and D.D. Smith (1978). Predicting rainfall erosion losses: A guide to conservation planning. *USDA Agricultural Handbook* #537, Washington, D.C.

Zaruba, Q., and V. Mencl. 1969. *Landslides and Their Control.* New York and Prague: Elsevier and Academia.

3 Influence of Vegetation on Hillside Stability

3.1 INTRODUCTION

Vegetation affects both the surficial and mass stability of slopes in significant and important ways. Various hydromechanical influences of vegetation, including methods for predicting and quantifying their magnitude and importance on stability, are described in this chapter. The stabilizing or protective benefits of vegetation depend on the type of vegetation and the type of slope-degradation process. In the case of mass stability, the protective benefits of woody vegetation range from mechanical reinforcement and restraint by the roots and stems to modification of slope hydrology as a result of soil moisture extraction via evapotranspiration.

The loss or removal of slope vegetation can result in either increased rates of erosion or higher frequencies of slope failure. This cause-and-effect relationship can be demonstrated convincingly as a result of many investigations reported in the technical literature (Gray and Sotir, 1996; Sidle et al., 1985; Swanston, 1974; O'Loughlin, 1974).

For the most part, vegetation has a beneficial influence on the stability of slopes; however, it can occasionally affect stability adversely or have other undesirable impacts—e.g., obstruction of views, hindrance of slope inspection, or interference of flood-fighting operations on levees. A number of strategies and techniques are described at the end of this chapter to maximize benefits and to minimize liabilities of plants. These include such procedures as the proper selection and placement of vegetation and site preparation and grading in addition to management techniques, such as pruning and coppicing.

The right choice of plant materials is critical. A tight, dense cover of grass or herbaceous vegetation, for example, provides one of the best protections against surficial rainfall and wind erosion. Conversely, deep-rooted, woody vegetation is more effective for mitigating or preventing shallow, mass-stability failures. In a sense, soil bioengineering and biotechnical methods also can be viewed as strategies or procedures for minimizing the liabilities of vegetation while capitalizing on its benefits.

3.2   INFLUENCE ON SURFICIAL EROSION

3.2.1   Stabilizing Functions

Vegetation plays an extremely important role in controlling rainfall erosion. Soil losses due to rainfall erosion can be decreased a hundredfold (USDA Soil Conservation Service, 1978) by maintaining a dense cover of sod, grasses, or herbaceous vegetation. The beneficial effects of herbaceous vegetation and grasses in preventing rainfall erosion are tabulated below:

- *Interception:* Foliage and plant residues absorb rainfall energy and prevent soil detachment by raindrop splash.
- *Restraint:* Root systems physically bind or restrain soil particles while aboveground portions filter out sediment from runoff.
- *Retardation:* Stems and foliage increase surface roughness and slow velocity of runoff.
- *Infiltration:* Plants and their residues help to maintain soil porosity and permeability, thereby delaying onset of runoff.

In the case of surficial erosion, herbaceous vegetation and grasses are more effective than woody vegetation, because they provide a dense, ground cover.

3.2.2   Vegetation Cover Factor

A good gauge of the influence of vegetation in preventing soil erosion can be obtained by examining the Universal Soil Loss Equation (USLE). The annual soil loss from a site is predicted according to the following relationship:

$$A = R \cdot K \cdot LS \cdot C \cdot P \qquad (3.1)$$

where A = computed soil loss (e.g., tons) per acre for a given storm period or time interval;

R = rainfall factor;
K = soil erodibility value;
L = slope-length factor;
S = steepness factor;
C = vegetation factor;
P = erosion control—practice factor.

The USLE provides a simple, straightforward method of estimating soil losses, and it provides an idea of the range of variability of each of the parameters, their relative importance in affecting erosion, and the extent to which each can be changed or managed to limit soil losses. The climatic (R), topographic (*LS*), and erodibility (*K*) factors only vary within one order of magnitude. The vegetation or cover (*C*) factor, on the other hand, can vary over several orders of magnitude as shown in Table 3.1.

Moreover, unlike the other factors, the cover (*C*) factor can be radically decreased by the selection, method of installation, and maintenance of a particular cover system. Factor C values tend to change with time following certain types of surface treatment, such as mulching, seeding, and transplanting. For example, factor C values for grass may decrease from 1.0 (for fallow, bare ground) to about 0.001 between time of initial seeding and full establishment with a dense, grass sod.

3.2.3 Recommended Vegetation

Under normal conditions, a dense cover of grass or herbaceous vegetation provides the best protection against surficial rainfall and wind erosion. A grass cover can be established by either seeding or sodding. Seed mixtures normally include grasses that germinate rapidly—such as rye or annual grass—to provide immediate short-term protection and slower-growing perennial grasses that take more time to establish but provide long-term protection. The optimum seed mix depends on soil, site, and climatic conditions. A horticulturist

TABLE 3.1   Cover Index Factor (C) for Different Ground Cover Conditions

Type of Cover		Factor C	Percent(*) Effectiveness
None (fallow ground)		1.0	0.0
Temporary seedings (90% stand):			
Ryegrass (perennial type)		0.05	95
Ryegrass (annuals)		0.1	90
Small grain		0.05	95
Millet or Sudan grass		0.05	95
Field bromegrass		0.03	97
Permanent seedings (90% stand):		0.01	99
Sod (laid immediately)		0.01	99
Mulch:			
Hay rate of appl., tons/ac:			
"	0.5	0.25	75
"	1.0	0.13	87
"	1.5	0.17	93
"	2.0	0.02	98
Small grain straw	2.0	0.02	98
Wood chips	6.0	0.06	94
Wood cellulose	1.5	0.10	90
Fiberglass	1.5	0.05	95

*Percent soil-loss reduction as compared with fallow ground.

Source: From USDA Soil Conservation Service (1978).

familiar with local conditions should be consulted for recommendations. Site preparation, mulching, and fertilization may also be required to insure germination and establishment.

3.3   INFLUENCE ON MASS STABILITY

3.3.1   Hydromechanical Effects

The protective role of vegetation on the stability of slopes has gained increasing recognition (Coppin et al., 1990). Greenway (1987) provides a good summary of the hydromechanical influences of vegetation as related to mass stability. These influences are depicted schematically in Figure 3.2 and tabulated in Table 3.2 according to whether they exert a beneficial or adverse effect.

Some mechanisms—e.g., surcharge—may be either beneficial or adverse depending on soil and slope conditions, as discussed later in this chapter. The principal destabilizing mechanism is probably from windthrowing, which can cause local instability in a slope. Depletion of soil moisture by vegetation is a two-edged sword: on the one hand, evapotranspiration leads to lower pore water pressures in a slope; on the other, soil moisture depletion can accentuate desiccation cracking, which in turn leads to higher infiltration capacity.

3.3.2   Beneficial Effects

The main beneficial effects of woody vegetation on the mass stability of slopes are listed below:

- *Root Reinforcement:* Roots mechanically reinforce a soil by transfer of shear stress in the soil to tensile resistance in the roots.
- *Soil-Moisture Depletion:* Evapotranspiration and interception in the foliage can limit buildup of positive pore water pressure.
- *Buttressing and Arching:* Anchored and embedded stems can act as buttress piles or arch abutments to counteract downslope shear forces.
- *Surcharge:* Weight of vegetation can in certain instances increase stability via increased confining (normal) stress on the failure surface.

The most obvious way in which woody vegetation enhances mass stability is via root reinforcement. Extensive laboratory studies (Gray and Ohashi, 1983; Maher and Gray, 1990) on fiber-reinforced sands indicate that small amounts of fiber can provide substantial increases in shear strength. These findings have been corroborated by laboratory and field tests on root-permeated soils (Endo and Tsuruta, 1969; Waldron, 1977; Ziemer, 1981; Riestenberg, 1983, 1994; and Nilaweera, 1994).

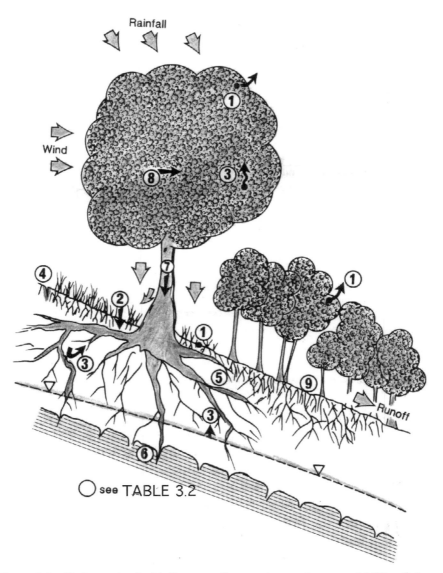

Figure 3.1 Hydromechanical influences of vegetation on the mass stability of slopes (from Greenway, 1987).

The soil buttressing and arching action associated with roots and the stems and trunks of woody vegetation are also important components of slope stabilization. In addition, evapotranspiration by vegetation can reduce pore water pressures within the soil mantle on natural slopes, promoting stability (Brenner, 1973). These beneficial effects have been analyzed in detail by Greenway

TABLE 3.2   Hydromechanical Influences of Vegetation on the Mass Stability of Slopes

Hydrologic Mechansims	Influence
1. Foliage intercepts rainfall, causing absorptive and evaporative losses that reduce rainfall availability for infiltration.	Beneficial
2. Roots and stems increase the roughness of ground surface and permeability of soil, leading to increased infiltration capacity.	Adverse
3. Roots extract moisture from the soil; moisture that is lost to the atmosphere via transpiration leads to lower pore water pressures.	Beneficial
4. Depletion of soil moisture may accentuate desiccation cracking in the soil, resulting in higher infiltration capacity.	Adverse
Mechanical Mechanisms	Influence
5. Roots reinforce the soil, increasing shear strength.	Beneficial
6. Tree roots may anchor into firm strata, providing support to the upslope mantle through buttressing and arching.	Beneficial
7. Weight of trees surcharges the slope, increasing both normal and downhill force components	Adverse/Beneficial
8. Vegetation exposed to wind transmits dynamic forces into slope with danger of windthrowing and toppling.	Adverse
9. Roots bind soil particles at ground surface and increase surface roughness, thereby reducing susceptibility to erosion.	Beneficial

Source: Adapted from Greenway (1987).

(1987) and Gray and Sotir (1996). Important findings are reviewed and summarized in succeeding sections of this chapter.

3.3.3 Detrimental Effects

The primary detrimental influence on mass stability associated with woody vegetation appears to be the concern about external loading and the danger of overturning or uprooting in high winds or currents (Nolan, 1981; Tschantz and Weaver, 1988). If a significantly sized root ball is unearthed during uprooting, it could reduce the stability of a cross section, depending upon a tree's position on the slope. This problem is likely to be more critical for large trees growing on relatively small dams, levees, or streambanks. With regard to external loading, levee-embankment slopes are generally shallow enough that the main component of the overburden weight may act perpendicular to, rather than parallel to, the failure surface, thereby increasing stability. However, the location of trees on the embankment must be considered in any slope-stability analysis in order to ascertain the extent to which their weight might affect the balance of forces.

The problem of surcharge and windthrowing can be eliminated by proper plant selection or, alternatively, by the practice of coppicing. Planting shrubs or small trees, for example, with a large root-to-shoot ratio increases the belowground relative to the aboveground biomass, thereby minimizing problems associated with surcharge or windthrowing. Coppicing, a type of pruning procedure, also allows one to create an unobstructed view (a frequent reason for tree removal on slopes) by effectively reducing tree heights while retaining all the benefits provided by a tree's living root system. These mitigating practices are described in greater detail at the end of the chapter (see Section 3.6).

3.4   ROOT MORPHOLOGY AND STRENGTH

3.4.1   Introduction

Woody vegetation affects shallow-mass stability mainly by increasing the shear strength of soil via root reinforcement. The value of the root system in this regard will depend upon the strength and interface properties of the roots themselves and on the concentration, branching characteristics, and spatial distribution of roots in the ground. Root strength and architecture, in turn, are governed by the type of plant and by local soil and site conditions. Gray and Sotir (1996) provide details and additional information about the different characteristics of root systems

3.4.2   Depth and Distribution of Root Systems

Deeply penetrating vertical taproots and sinker roots provide the main contribution to the stability of slopes vis-à-vis resistance to shallow sliding. Mechanical restraint against sliding only extends as far as the depth of root penetration. In addition, the roots must penetrate across the failure surface to have a significant effect. The influence of root reinforcement and restraint for different slope stratigraphies and conditions is summarized in Figure 3.2. The most effective restraint is provided where roots penetrate across the soil mantle into fractures or fissures in the underlying bedrock (Case B) or where roots penetrate into a residual soil or transition zone whose density and shear strength increase with depth (Case C).

Because of oxygen requirements, the roots of most trees tend to concentrate near the surface. As a rough rule of thumb, the mechanical reinforcing or restraining influence of roots on a slope is probably limited to a zone about 5 feet (1.5 m) from the surface. Studies by Patric et al. (1965) in a loblolly pine plantation showed that 80 to 90 percent of the roots in their test plots were concentrated in the first 3 feet (0.9 m). The bulk of the near-surface roots were laterals; in contrast, roots below 3 feet were generally vertically oriented.

Root-Area Ratios: Root-area ratios (RAR) were measured as a function of depth in a sandy levee along the Sacramento River in California (Shields and Gray, 1993) for a variety of woody plant species. Root-area ratio

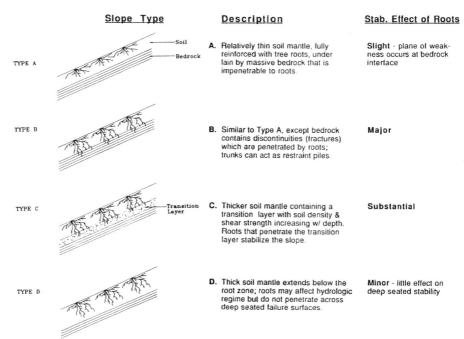

Slope Type	Description	Stab. Effect of Roots
TYPE A	**A.** Relatively thin soil mantle, fully reinforced with tree roots, under lain by massive bedrock that is impenetrable to roots.	**Slight** - plane of weakness occurs at bedrock interface
TYPE B	**B.** Similar to Type A, except bedrock contains discontinuities (fractures) which are penetrated by roots; trunks can act as restraint piles.	**Major**
TYPE C	**C.** Thicker soil mantle containing a transition layer with soil density & shear strength increasing w/ depth. Roots that penetrate the transition layer stabilize the slope.	**Substantial**
TYPE D	**D.** Thick soil mantle extends below the root zone; roots may affect hydrologic regime but do not penetrate across deep seated failure surfaces.	**Minor** - little effect on deep seated stability

Figure 3.2 Influence of slope stratigraphy on the stabilizing effect of roots against slope failure (adapted from Tsukamoto and Kusuba, 1984).

refers to the fraction of the total cross-sectional area of a soil that is occupied by roots. Few roots were encountered below a depth of 4 feet in either transect.

The stabilizing effect of roots is lowest when there is little or no penetration across the shear interface (Figure 3.2, Case A and D). However, even in these cases, lateral roots can play an important role by maintaining the continuity of a root-permeated soil mantle on a slope.

Root Spread: Tree roots can spread out for considerable distances; in one reported instance (Kozlowski, 1971), roots of poplars growing in a sandy soil extended out 65 meters. The extent of root spread is normally reported in relative multiples of the tree height or crown radius. Kozlowski (1971) cites 10-year-old pine trees growing on sandy soil with a root spread about seven times the average height of the trees, probably an extreme case. More typical of root systems was the case of fruit trees growing on clay that had roots extending 1.5 times the crown radius. Similar trees growing on loam extended 2 times, and those on sand up to 3 times the crown radius. A useful rule of thumb is that a root system will spread out a distance at least equal to the 1.5 times the radius of the crown. The hydraulic influence of a tree—i.e., significant soil moisture reductions caused by evapotranspiration—can be felt to distance of at least 1 times the tree height. These findings have implications with regard to both slope stability and safe placement of structures adjacent to trees growing on compressible soils.

Factors Affecting Root Development: Root development and structure are affected initially by genetic disposition but, ultimately, are governed more by environmental and edaphic conditions (see Sutton, 1971). Henderson et al. (1983) have noted that root systems tend to grow wide and deep in well-drained soils as opposed to developing a flat, platelike structure in a surface soil underlain by a more dense or rocky substratum.

The degree to which roots are able to penetrate underlying bedrock depends to a large extent on the nature and extent of discontinuities (i.e., joints and fractures) in the bedrock. Trees growing in shallow, coarse-textured soils developed on granitic bedrock, for example, can develop sinker roots and taproots that penetrate into fissures and fractures in the underlying bedrock, as shown in Figure 3.3. The overlying soil developed on granitic bedrock is often coarse and incapable of holding much moisture; consequently, roots seek out water in the fractures and fissures in the underlying bedrock. This adaptation in turn insures that the trees

Figure 3.3 Exposed root system of a western yellow pine (*Pinus ponderosa*) showing development of extensive, vertical root system that penetrates into underlying fractures in the bedrock.

will be well anchored to the slope and help to restrain movement of the soil mantle by a combination of buttressing and arching action (Gray and Sotir, 1996).

Root Structure and Distribution—Experimental Methods: Various methods for determining root structure and distribution are described in detail elsewhere (Bohm, 1979; Gray and Sotir, 1996; Gray and Barker, 2004). Root-area ratios or root biomass concentration as a function of depth are required in order to estimate rooting contributions to soil shear strength. This ratio varies spatially in three dimensions, as shown schematically in Figure 3.4. The area ratio of greatest interest coincides with the critical sliding surface. Normally, this surface is oriented parallel to the slope.

One approach is simply to recover large samples of root-permeated soil from various depths and measure the root biomass per unit volume at each depth by sieving the soil, recovering, and weighing the roots. Root biomass per unit volume can be converted to an equivalent root-area ratio, if the unit weight or density of the roots is known. Alternatively, root-area ratios can be measured directly in an excavated trench

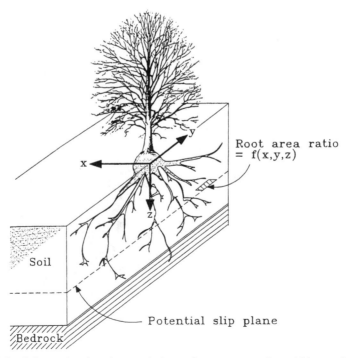

Figure 3.4 Schematic showing variation of root-area ratio within a slope (from Greenway, 1987).

using the "profile wall" method (Bohm, 1979; Gray and Sotir, 1996). In this method, the roots exposed in the vertical side of the trench are carefully mapped by means of a gridded, acetate overlay.

Perhaps the best way to ascertain root architecture and distribution is to exhume or uncover the roots in situ. This can be accomplished by cutting and removing the trunk and then exposing the roots by removing the surrounding soil. The soil can be removed by hydraulic washing or, alternatively, by using an air-jet device known as an "Airspade." Soil particles surrounding roots are dislodged and removed by means of high-pressure air jets at the tip of the Airspade. Virtually, the entire root system of a tree can be exhumed more or less intact and studied in this manner, as shown in Figure 3.5.

3.4.3 Root Strength

Factors Affecting Strength: Wide variations in tensile strength of roots have been reported in the technical literature, depending on species and such site factors as growing environment, season, root diameter, and orientation. Greenway (1987) has compiled an excellent review of root strength and factors affecting it. With regard to the influence of seasonal effects, Hathaway and Penny (197) reported that variations in specific gravity and lignin-cellulose ratio within poplar and willow roots produced seasonal fluctuations in tensile strength. It is also important to

Figure 3.5 Root architecture and distribution of white oak exhumed in situ, using an "Airspade" (from Gray and Barker, 2004).

recognize that root tensile strength is affected as much by differences in size (diameter) as by species. Several investigators (Turnanina, 1965; Burroughs and Thomas, 1977; and Nilaweera, 1994) have reported a decrease in root tensile strength with increasing size (diameter). Roots are no different in this regard than fibers of other materials, which exhibit a similar trend.

Ranges in Root Tensile Strength and Modulus: Root tensile strengths have been measured by a number of different investigators. Nominal tensile strengths reported in the technical literature are summarized in Table 3.3 for selected shrub and tree species. Tensile strengths vary significantly with diameter and method of testing (e.g., in a moist or air dry state). Accordingly, the values listed in Table 3.3 should be considered

TABLE 3.3   Nominal Tensile Strength of Selected Tree and Shrub Species

Species	Common Name	Mean Tensile Strength (MPa)
TREE SPECIES:		
Abies concolor	Colorado white fir	11
Acacia confusa	Acacia	11
Alnus firma var. *multinervis*	Alder	52
Alnus incana	Alder	32
Alnus japonica	Japanese alder	42
Betula pendula	European white birch	38
Nothofagus fusca	Red beech	32
Picea sitchensis	Sitka spruce	16
Picea abies	European spruce	28
Pinus densiflora	Japanese red pine	33
Pinus lambertiana	Sugar pine	10
Pinus radiata	Monterey pine	18
Populus deltoides	Poplar	37
Populus euramericana 1488	American poplar	33
Pseudotsuga mensieii	Douglas fir (Pacific coast)	55
Pseudotsuga mensieii	Douglas fir (Rocky mountains)	19
Quercus robur	Oak	20
Sambucus callicarpa	Pacific red elder	19
Salix fragilis	Crack willow	18
Salix helvetica	Willow	14
Salix matsudana	Willow	36
Salix purpurea (Booth)	Purple willow	37
Tilia cordata	Linden	26
Tsuga heterophylla	Western hemlock	20
SHRUB SPECIES:		
Castanopsis chrysophylla	Golden chinkapin	18
Ceanothus velutinus	Ceanothus	21
Cytisus scoparius	Scotch broom	33
Lespedeza bicolor	Scrub lespedeza	71
Vaccinium spp.	Huckleberry	16

Source: Adapted from Schiechtl (1980).

only as rough or approximate averages. Nevertheless, some interesting trends can be observed in the tabulated strength values. Tensile strengths can approach 70 MPa but appear to lie in the range of 10 to 40 MPa for most species. Conifers as a group tend to have lower root strengths than deciduous trees. Shrubs appear to have root tensile strengths at least comparable to that of trees. This is an important finding because equivalent reinforcement can be supplied by shrubs at shallow depths without the concomitant liabilities of trees, resulting from their greater weight and tendency for windthrowing.

3.5   ROOT AND FIBER SOIL REINFORCEMENT

The mechanical or reinforcing effect of plant roots on the stability of slopes can be described and accounted for in a systematic manner. Root fibers reinforce a soil by transfer of shear stress in the soil matrix to tensile resistance in the fiber inclusions. When shear occurs, the fiber is deformed or stretched. This deformation causes the fiber to elongate, provided there is sufficient interface friction and confining stress to lock the fiber in place and prevent slip or pullout. As the fiber elongates, it mobilizes tensile resistance in the fiber. The component of this tension, tangential to the shear zone, directly resists shear, while the normal component increases the confining stress on the shear plane.

3.5.1   Force-Equilibrium Models

Simple force-equilibrium models are useful for identifying parameters that affect root reinforcement and predicting the amount of strength increase from the presence of fibers in a soil. These models and laboratory verification tests are described in detail elsewhere (Gray and Barker, 2004; Waldron, 1977; Waldron and Dakessian, 1981; Wu et al., 1979; Wu et al., 1988a, 1988b; Shewbridge and Sitar, 1989, 1990). Three different response scenarios are possible during shearing of a fiber-reinforced soil composite, namely, fibers break, stretch, or slip. The predictive equations and their corresponding assumptions for these response scenarios are presented elsewhere (Gray and Barker, 2004; Gray and Sotir, 1996).

3.5.2   In Situ Direct-Shear Tests

Ziemer (1981) conducted in situ direct-shear tests on sands permeated with live roots of pine trees (*Pinus contorta*). The largest roots exposed in this shear cross section were under 17 millimeters. He observed an approximately linear increase in shear strength with increasing root biomass, as shown in Figure 3.6.

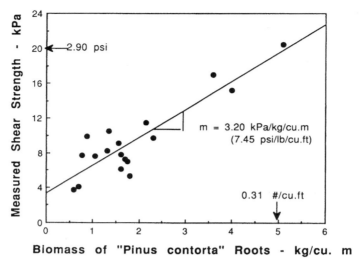

Figure 3.6   Results for in situ direct-shear tests on sand permeated with pine roots (from Ziemer, 1981).

Results of Ziemer's in situ tests in a sand, using pine roots. and laboratory tests performed by Gray and Ohashi (1979) on a sand, using reed fibers, are compared in Table 3.4. The shear-strength increase response was very similar in both cases; the shear-strength increase per unit fiber concentration ranged from 3.2 to 3.7 kPa/kg of root/m^3 soil (7.4 to 8.7 psi/lb of root/cf soil). This translates into an average value of 3.2 psi/percent RAR [assuming a root unit weight of 40 pounds per cubic foot (pcf)].

These unit shear-strength increases can be used to obtain order of magnitude estimates for rooting contribution to shear strength in stability analyses (see Sec. 3.5.3). All that is required is a measure of the root biomass at the depth of interest using the techniques described in Section 3.4.2, for example, exhumation or profile-wall method.

3.5.3   Stability Analyses

Conventional slope-stability analyses (see 2.9, Slope Stability Predictions) can be used to ascertain the influence of root reinforcement on mass stability. The "infinite slope" model is appropriate for analyzing shallow slides in which the failure surface is planar and parallel to the slope over most of its length (see 2.9.4, Translational Slope Failures). The infinite slope model is generally applicable for sandy slopes that are subject to shallow sloughing. Cohesionless soils are weakest near the surface, where confining stress and, hence, shear resistance is lowest. The infinite slope model also is valid for analyzing

TABLE 3.4   Summary of Root and Fiber Contributions to Soil Shear Strength of Sands Based on Laboratory and In Situ Tests

Fiber or Root System	Max Fiber or Root Biomass Conc.		Shear Str. Incr. Per Unit Fiber Conc.	
	Area Ratio (%)	Wt. Conc. lbs root cf soil	PSI lb/cf	kPa kg/cu m
IN SITU TESTS ON ROOT-PERMEATED SAND				
Tree roots (*pinus contorta*)	0.78	0.31	7.4	3.2
Vertical shear surface, coastal sand				
Live roots <17 mm diameter				
In situ direct shear test				
LAB TESTS ON FIBER-PERMEATED SAND:				
Reed fibers (*phragmites communis*)	1.70	0.68	8.7	3.7
Natural fibers: diameter = 2.0 mm				
Uniform sand				
Direct shear test				
		AVERAGES	8.1	3.5

slopes with a residual soil mantle underlain by a shallow bedrock contact—the most likely sliding surface.

The influence of groundwater must also be considered in stability analyses. Water plays an important role in both shallow mass wasting and surficial erosional processes, either by subsurface seepage or by overland flow. The presence of groundwater in a slope causes pore water pressures to develop that can increase shear stress and decrease shear resistance. In addition, groundwater flow or seepage in a slope can exert destabilizing stresses, depending on the direction of flow. Groundwater flow or seepage that exits at or flows toward the free face of a slope presents a particular threat.

The factor of safety of an infinite slope that is subject to seepage can be computed as a function of vertical depth (H) and seepage direction (θ) with respect to a horizontal plane (see Figure 3.7). The factor of safety for this case is given by the following relationship:

$$F = A[\text{Tan}(\phi)/\text{Tan}(\beta)] + B[(c + c_R)/\gamma H] \tag{3.2}$$

$$A = [1 - r_u/\text{Cos}^2(\beta)] \tag{3.3}$$

$$B = [1/\text{Cos}^2(\beta)\text{Tan}(\beta)] \tag{3.4}$$

$$r_u = [(\gamma_W/\gamma)\{1/(1 + \text{Tan}(\beta)\text{Tan}(\theta)\}] \tag{3.5}$$

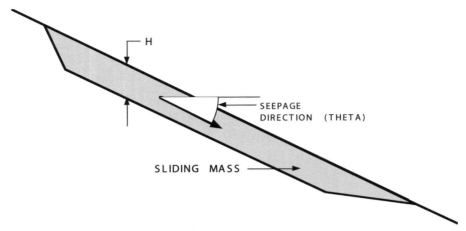

Figure 3.7   Infinite slope model for a cohesionless soil with seepage parallel to slope surface.

where β = slope angle;
ϕ = angle of internal friction;
θ = seepage angle (with respect to horizontal);
H = vertical depth below surface;
c = soil cohesion;
c_R = root cohesion;
γ = soil density;
γ_w = density of water.

Stability without Roots: The infinite slope equation can be programmed into a spreadsheet and a factor of safety computed (Gray and Sotir, 1996; Gray and Barker, 2004) for different assumed values of soil friction (ϕ), soil cohesion (c), root cohesion (c_R), seepage direction (θ), depth (H), and slope angle (β). If the seepage angle is the same as the slope angle ($\theta = \beta$), then seepage occurs parallel to the slope. If the seepage angle is zero ($\theta = 0°$), then seepage occurs horizontally toward the face of the slope. This type of horizontally exiting, or "daylighting," seepage results in the lowest safety factors. Seepage parallel to the slope or, worse yet, emerging from the slope tends to cause sloughing and instability at the face. On the other hand, when seepage is vertically down ($\theta = 90°$), this condition yields a relatively high factor of safety.

Stability with Roots: The factor of safety calculations for slopes without roots ($c_R = 0$) can be repeated, taking into account the presence of roots. A root-reinforcement factor determined from either laboratory or in situ shear-strength tests (see Table 3.4) can be used along with an estimate of the root biomass concentration (or root-area ratio) to compute a shear strength increase (or root cohesion) versus depth relationship.

Methods of determining the root biomass or area ratio (RAR) as a function of depth were described previously (see 3.4.2). For illustrative purposes, a root-reinforcement factor of 3.2 psi/percent RAR can be used based on results of shear-strength tests on root-permeated sands reported in Table 3.4. A root-area ratio ranging from 0.119 percent at a depth of 0.5 ft to 0.013 percent at a depth of 4.0 ft can be assumed, based on results from profile wall tests in a vegetated, sandy levee (Shields and Gray, 1993). The product of these RARs times the root-reinforcement factor yields the shear-strength increase from the presence of roots (or root cohesion) at successive depths.

Root reinforcement greatly enhances stability at shallow depths (<3 ft) even for the case of emergent or lateral seepage ($\theta > \beta$). At greater depths (>3 ft), the effect of root cohesion diminishes and the curves approach the same values for the case of an unreinforced, cohesionless slope. This follows both because there are fewer roots at greater depths and because the contribution of cohesion to the total shear resistance decreases with increasing depth.

3.6   GUIDELINES FOR MAXIMIZING BENEFITS OF VEGETATION

3.6.1   General Observations

As discussed previously, vegetation generally has a beneficial influence on mass stability of slopes; but it can also have detrimental or adverse effects as well (see to Table 3.1). Fortunately, several strategies and procedures can be adopted to maximize the benefits of vegetation while minimizing its liabilities. These strategies include: (1) selection of the appropriate species for particular site conditions and stabilization objectives, (2) proper placement or location of vegetation, (3) grading and site preparation, (4) attention to compaction procedures, and (5) management of the vegetation to mitigate undesirable characteristics or problems. These strategies and procedures are summarized in Table 3.5.

3.6.2   Selection Strategies

Vegetation should be selected for desired stabilization objectives and be compatible with soil and site conditions. The latter includes consideration of soil type, water availability, nutrient status, and soil pH, climate, possible browsing pressure, regulations governing the use of exotic or nonnative species, and so forth. Certain types of plants are intrinsically better suited than others for specific stabilization objectives. Woody vegetation is stronger and deeper rooted than herbaceous plants and grasses and provides greater mechanical

TABLE 3.5   Strategies for Maximizing Benefits of Vegetation

1. *SPECIES SELECTION*
   • Compatibility with soil and site conditions
   • Functional suitability (grasses vs. woody plants)

2. *PLACEMENT STRATEGIES*
   • Locate according to size and height on slopes
   • Group according to "landform revegetation" principles
   • Maintain viewing inspection "alleys"

3. *GRADING AND SITE PREPARATION*
   • Conventional vs. "landform" grading
   • Surface modification (decompacting and ripping)
   • Soil blending and artificial gradation
   • Soil amendments
   • Surface and subsurface drainage

4. *COMPACTION PROCEDURES AND CONTROL*
   • Optimizing compaction to balance engineering requirements vs. plant needs

5. *MANAGEMENT AND MAINTENANCE STRATEGIES*
   • Watering protocols
   • Weeding and herbicides
   • Fire management
   • Protection from grazing and browsing
   • Fertilizers and soil amendments
   • Pruning (coppicing and pollarding)
   • Control of pedestrian traffic

reinforcement and buttressing action at depth. Accordingly, woody plants are superior for mass stability. Grasses and herbaceous vegetation, on the other hand, grow close to the surface and provide a tight, dense ground cover. They tend to be superior, therefore, in intercepting rainfall and preventing surficial erosion. Shrubs are neither as deeply rooted as trees nor can they be expected to provide as much buttressing restraint as trees. On the other hand, shrubs are more flexible, have less aboveground biomass, and exert less surcharge on a slope. They may be preferable, accordingly in riverbank and levee stabilization, where these attributes are advantageous. The characteristics and relative advantages and disadvantages of different plant types for various applications are summarized in Table 3.6.

3.6.3   Placement Strategies

Several different placement or locational strategies can be invoked to maximize the utility of slope plantings and minimize possible problems. One of the main objections raised to vegetation on slopes is that it obstructs views and hinders access. These objections have been raised by both homeowners living on hillsides and inspectors examining river levees. These problems can be addressed by pruning and coppicing techniques (described in the next section). They can also be addressed by placement of vegetation on a slope

TABLE 3.6   Characteristics of Different Plant Types and Suitability for Various Applications

Type	Advantages	Disadvantages
1. Grasses	Good dense surface coverage. Wide range of tolerances. Quick to establish. Relatively inexpensive.	Shallow rooting. Regular maintenance required. Competes with woody vegetation.
Reeds and sedges	Establish well on riverbanks, etc. Quick growing.	Hand planting is expensive. Difficult to obtain.
Dune grasses	Trap drifting sand and build protective dunes. Thrive in droughty and harsh conditions.	Shade intolerant and out competed by other plants on relatively stable sites.
2. Herbs	Deeper rooting. Attractive in grass sward.	Seed expensive. Sometimes difficult to establish. Many species die back in winter.
Legumes	Cheap to establish. Fix nitrogen. Mix well with grass.	Not tolerant of difficult sites.
3. Shrubs	Hardy and fairly cheap. Many species can be seeded. Substantial ground cover. Deeper rooting. Low main tenance. Many evergreen species.	More expensive to plant. Some times difficult to establish.
4. Trees	Substantial rooting. Some can be seeded. No maintenance once established.	Long time to establish. Slowgrowing. Relatively expensive. Ineffective for surface erosion.
Willows and dogwoods	Root easily from cuttings. Applicable to soil bioengineering treatments. Relatively easy to establish.	Care required in selecting correct type. Cuttings must be placed during dormant season.

according to its height and the shape or density of the foliage crown (see Figure 3.8). Smaller shrubs should be grown near the top of slope and larger trees placed near the bottom. This simple procedure will not only improve views from the top but also eliminate weight from the top of the slope and put maximum buttressing restraint and reinforcement near the base where it is most needed.

3.6.4   Grading and Site Preparation

Landform Grading: A key strategy is to locate vegetation in conformance with "landform grading" practices, as discussed later in Chapters 7, 8, and 9. Revegetation in conjunction with landform grading entails planting vegetation in patterns that occur in nature as opposed to specifying

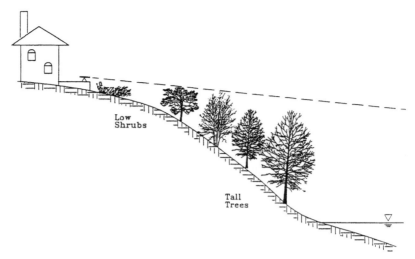

Figure 3.8 Placement of vegetation on slope according to size and form to maximize stabilizing effect and to minimize view obstruction.

either uniform or random coverage. Trees and large shrubs tend to require more moisture, and they are also better at stabilizing against shallow-slope failures than herbaceous vegetation. Accordingly, trees should be clustered in swales and valleys in a slope where runoff tends to concentrate and evaporation is minimized. Conversely, convex slope areas should be planted with more drought-tolerant grasses and herbaceous vegetation. Irrigation needs can be reduced as much as 40 percent by careful control of drainage pattern on a slope and selection and placement of appropriate plantings for different areas.

Surface Modification: Better plant establishment and initial growth can be achieved if the soil surface is mechanically modified either by contour furrowing, scarification, disking, track walking, pitting, ripping, chiseling, or land imprinting. These treatments transform smooth, sealed soil surfaces with low infiltration rates into microrough, macroporous surfaces that are better able to exchange water and air rapidly across the air-earth interface. They tend to loosen the surface layers and provide small indentations where seed and water can collect, thus aiding germination and establishment of plants. Descriptions and specifications for these various mechanical treatments have been issued by the Natural Resources Conservation Service (USDA, 2000). An example of the use of ripping to improve vegetative establishment as part of a watershed restoration program in Redwood National Park is shown in Figure 3.9. Track walking also helps to anchor mulch that is applied to the surface. Scarification, ripping, etc, loosen and initially increase the risk of some

(*a*)

(*b*)

Figure 3.9 Soil decompaction on abandoned roadbed, Redwood National Park, California: (*a*) use of root plow to rip and loosen soil and (*b*) revegetation of decompacted road surface.

surficial erosion and sloughing, but this initial risk must be weighed against better long-term protection that is afforded by a vigorous, well-established vegetative cover. Furthermore, initial surficial erosion can be minimized or controlled by the use of hydraulically applied soil binders, fiber mulches, erosion-control blankets, and soil bioengineering treatments such as the use of live fascines (Gray and Sotir, 1996).

Soil Blending and Artificial Gradation: Soil blending entails adding coarse, noncohesive particles to a soil to improve water transmission properties, allow densification, and still permit good plant growth. Arborists who have to confront the contradictory demands of plant growth versus engineering stability have often resorted to this approach. Street trees are usually grown in soil that must be compacted to a high degree to provide stability to adjoining sidewalks, roadways, and buildings. This same soil must be capable of accommodating growth, very often under less than ideal conditions. One way around this problem is to use a type of soil referred to as "structural" soil. This type of soil allows the granular portion to be compacted to a high relative density while still providing enough pore space to accommodate needed fines and plant roots in the intergranular voids. Mitchell (1993) describes ways of actually computing the relative proportions of granular solids and fines (clay and silt) to achieve this goal.

Another version of structural soil (LASN, 2001) that has been used in street tree plantings consists of 4 to 5 parts crushed rock mixed with peat soil. The rock varies in size between ½ and 1½ inches in diameter. The soil is 25 percent silt or clay, 25 percent organic matter, and 50 percent fine sand. A soil stabilizer (plant-derived glue or hydrogel) is mixed in so that the soil adheres to the crushed rock. The ratio is 30 grams of soil stabilizer per 100 kilograms of soil and 500 kilograms of crushed rock. Preparation of this structural soil involves spreading out the rock and mixing in the soil stabilizer and soil into the rock.

Surface Amendments and Treatments: Specially formulated soil mixes can also be added atop the surface not only to improve plant growth and establishment but also to minimize runoff and erosion. A good example is the recent development of an organic based soil (Durant, 2001) that mimics the texture and microbiology of native topsoil at depths of 30 centimeters to 91 centimeters (12 to 36 in.) and remains stable on 1V:1H slopes or steeper. This soil mix reportedly does not erode, rill, or slump during test simulations with intense rainfall on steep slopes. The organic soil mix contains the necessary bacteria, fungi, and mycorrhizae that promote vigorous plant growth and that help to modify the underlying compacted, mineral soil and make it more hospitable to plant growth in the long run.

3.6.5 Optimizing Compaction

Purpose of Compaction: A conflict appears to exist between engineering requirements to compact soil to a high density to improve its engineering properties (e.g., increased strength and decreased compressibility) and agronomic needs to maintain soil in a relatively loose condition to improve its ability to support vegetation. The objectives of compaction, from an engineering perspective, have frequently been obscured in a manner that makes accommodation with plant growth needs more difficult to achieve. Furthermore, vegetation can be grown successfully in compacted soil under less than ideal conditions provided certain limits and precautions are observed (Gray, 2003; Goldsmith et al., 2002).

Compaction can be defined as a process of densification due to the removal of air voids when external stress is applied to a soil. The purpose of compaction from an engineering viewpoint is NOT to increase soil density. An increase in soil density is a result or consequence of compaction—but not the goal. Density is used as a target in engineering soil-compaction specifications, but so too are molding water content, type of compaction, additives, and compactive effort. The purpose of compaction is to change the engineering properties of a soil in a desirable direction.

Relevant engineering or physical properties include strength, compressibility, volume stability (shrink-swell potential), hydraulic conductivity, and erodibility. In general, these properties change in a favorable direction with an increase in soil density. There are important exceptions, however, noted as follows: (1) swelling (heave) in clay soils tends to increase at higher densities and (2) strength can decrease significantly in clay soils compacted wet of optimum to high densities (a phenomenon referred to as "overcompaction").

The effects of soil compaction on soil strength, compressibility, hydraulic conductivity, and volume stability have been investigated thoroughly (Lambe and Whitman, 1969; Seed and Chan, 1959). In addition, a series of standardized testing procedures and methods for specifying compaction have been widely adopted. A typical compaction curve (or moisture-density relationship) is shown in Figure 3.10. Densities are normally expressed in terms of dry unit weight, namely, the dry weight of solids per unit volume. The dry density is related to the moist density or unit weight by the following equation:

$$\gamma_d = \frac{\gamma}{1 + \omega} \tag{3.6}$$

where γ = moist unit weight or density;
$\quad \gamma_d$ = dry unit weight or density;
$\quad \omega$ = water content (dry weight basis)

Higher densities are achieved when soil particles can pack closer together. The maximum density occurs at the so called "optimum water content," which varies with the type of soil and compactive effort. At

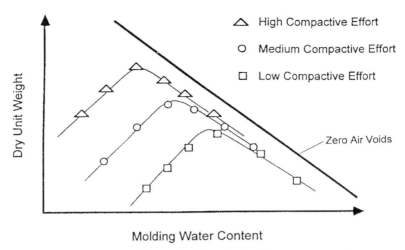

Figure 3.10   Typical moisture-density relationships or compaction curves.

optimum, the lubrication effect of the mix water allows soil particles to become more easily realigned during the compaction procedure, which results in the closer packing and higher density. At yet higher moisture content levels, the lubrication effect is offset by dilution, and dry density decreases. For any given textural composition of soil and compactive effort, there is a maximum dry density that can be achieved at the optimal moisture level, as shown schematically in Figure 3.10.

In general, compacted granular soils will have dry densities ranging from 115 to 135 pcf (18.1 to 21.2 kN/cubic meter) versus those of clayey to silty soils, which range from 85 to 115 pcf (13.4 to 18.1 kN/cubic meter). The corresponding optimum moisture content levels for a granular soil are on the order of 10 to 12 percent for granular soils and 15 to 20 percent for silty to clayey soils compacted by the Standard Proctor test. Maximum density does not represent a soil with no void space remaining; rather, it represents one where the tightest possible packing arrangement is achieved for the given compaction conditions. The point of 100-percent saturation is called the saturation line or "zero air voids" curve (see Figure 3.10). This condition cannot be reached unless the soil is completely saturated to begin with and is seldom, if ever, achieved during conventional compaction.

As a general rule in engineering practice, earthen fills that are part of site grading and not related to load bearing are specified to be compacted to 90 percent of Standard Proctor maximum dry density. Load-bearing soils and other specialized fill applications call for higher compaction levels, including compactions that exceed the values achieved by the Standard Proctor test. Typical compaction requirements (dry densities) for various engineering applications are summarized in Table 3.7.

Engineering Properties of Compacted Soil: Key variables affecting engineering soil properties during compaction include the following:

- Density (usually expressed in terms of "dry" unit weight)
- Water content (molding or mix water content during compaction)
- Compactive effort (energy input and unit volume of soil)
- Type of compaction (static, dynamic, or kneading)
- Additives (cement, lime, etc.)

The degree of saturation or water content of a clay soil at the time of compaction is perhaps the single most important variable that controls the engineering properties of the compacted material (Lambe, 1958). The influence of molding water content and compactive effort on hydraulic conductivity of compacted silty clay soil (Mitchell et. al., 1965) is shown in Table 3.8. Soils compacted at water contents less than optimum (dry of optimum) tend to have a relatively high hydraulic conductivity, whereas soils compacted at water contents greater than optimum (wet of optimum) tend to have a relatively low hydraulic conductivity. Higher molding water contents also greatly suppress hydraulic conductivity on the wet side of optimum, even offsetting the effect of decreased dry densities (or higher void ratios).

The results shown in Table 3.8 demonstrate that a soil compacted to

TABLE 3.7 Typical Compaction Requirements for Different Engineering Applications

Compacted Fill for:	Percent Modified Maximum Dry Density	Moisture Range about Opt. w/c
Roads		
Depth of 0 to 0.5 m	90 to 105	−2 to +2
Depth of. 0.5 m	90 to 95	−2 to +2
Small Earth Dam	90 to 95	−1 to +3
Large Earth Dam	95	−1 to +2
Railway Embankment	95	−2 to +2
Foundations for Structure	95	−2 to +2
Backfill Behind Walls	90	−2 to +2
Canal Linings	90	−2 to +2

Note: The relationship between Standard maximum and Modified maximum dry density is approximately as follows:

Sand: γ_{dmax} (Std.) $\approx 0.95 \gamma_{dmax}$(Mod.)
Clay: γ_{dmax} (Std.) $\approx 0.90 \gamma_{dmax}$(Mod.)

Source: From Hausmann (1990).

TABLE 3.8   Influence of Molding Water Content of a Silty Clay Compacted to Different Degrees of Compaction on the "Wet" and "Dry" Side of Optimum Water Content

Degree of Compact (% of Max Dry Density)	Dry Side Compaction		Wet Side Compaction	
	Water Content %	Hydraulic Conductivity cms/sec	Water Content %	Hydraulic Conductivity cms/sec
98	13.0	$0.5 \times 10\text{-}6$	16.0	$1.0 \times 10\text{-}8$
96	13.0	$1.0 \times 10\text{-}6$	17.0	$0.8 \times 10\text{-}8$
94	12.3	$2.0 \times 10\text{-}6$	18.5	$0.3 \times 10\text{-}8$
87	12.4	$7.2 \times 10\text{-}6$	22.5	$0.6 \times 10\text{-}8$

Note: Modified AASHTO compaction. Maximum dry density = 117.8 lbs/ft3; optimum water content = 15.0%.

the same degree of compaction on the wet side of optimum, using the same compaction method but at different *molding* water contents, can have radically different physical properties. This occurs because a soil compacted "wet" versus "dry" of optimum (see Figure 3.10) usually has a different texture or internal pore-size distribution. Soils compacted on the dry side of optimum water content tend to have a more open structure and greater distribution of larger pores. Accordingly, dry side compaction can result in hydraulic conductivities several orders of magnitude higher than wet side compaction, even when the soil is compacted to identical densities or degrees of compaction (see Table 3.8). This fact should always be kept in mind when assessing optimal compaction conditions to satisfy plant-growth needs vs. engineering requirements.

Influence of Soil Compaction on Plant Growth: Soil compaction can influence plant growth in a variety of ways, both good and bad. Agronomists generally recommend minimal soil compaction so as not to impede growth and development of crops and native plants. Soil must retain enough interconnected void space to allow storage and passage of air and water in the soil. Some degree of compaction is needed after planting or insertion of cuttings to close large voids and to provide suitable soil density for appropriate plant growth. Too much void space can lead to poor contact of a seed or cutting with the surrounding soil and subsequent desiccation.

The impacts of compaction have been studied extensively by agronomists who are concerned with the decline in soil productivity associated with modern agriculture, forestry practices, and the passage of equipment, which tend to compact soils over time. Goldsmith et al. (2001) provide a good review of these impacts on both conventional plantings and soil bioengineering installations. In general, findings show that high densities specified by engineers for mechanical strength tend to either reduce or effectively stop the development of roots. Depending on the plant species and the soil conditions, Goldsmith et al. (2001), cite evidence of limits to growth that include: (1) restriction in root growth, (2) severe reduction in length of all roots or primary root, (3) absence of root penetration of compacted soils. These authors conclude that a limiting or "threshold"-bulk density appears to exist for each soil type or texture above which plant growth is severely curtailed. They further suggest that these limiting densities may be used as a predictive or management tool.

Growth Limiting Bulk Densities for Plants: Several studies appear to support the concept of a growth-limiting bulk density (GLBD) that exists for a given soil texture or type. Daddow and Warrington (1983) computed GLBDs for 80 different soil textures using a regression equation. They next plotted the GLBDs on a USDA soil textural triangle to locate the growth-limiting isodensity lines, as shown in Figure 3.11.

Other researchers have tried to relate bulk density to factors such as root penetration, soil strength, and compaction. As noted previously, well-graded, noncohesive soils tend to reach higher maximum dry den-

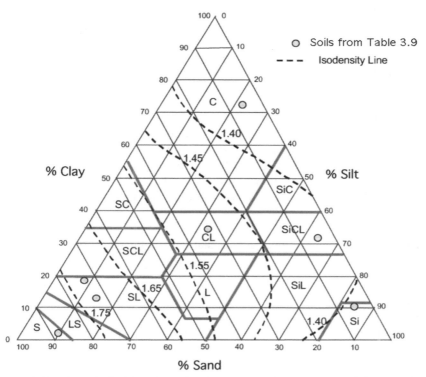

Figure 3.11 Growth-limiting bulk density textural triangle (adapted from Daddow and Warrington, 1983).

sities than cohesive soils. Additionally, noncohesive soils exhibit higher critical dry density than cohesive soils. Coppin and Richards (1990) concur that the critical dry density depends on the soil texture and suggests values of about 1.4 g/cubic cm (87 lb/cubic ft) for clay soils and 1.7 g/cubic cm (106 lb/cubic ft) for sandy soils. These threshold values are within the intervals presented in Table 3.9.

Clays contain more pore space than sandy soils, but they have a much smaller average pore size. The pore size distribution controls water transmission, not total porosity. Sandy soils have large pores, while clays have small pores, which transmit water slowly. On the other hand, soils with small pores retain and hold moisture more effectively.

Optimal conditions occur when there are enough large pores to transmit water readily but also enough small pores to retain and store water; therefore, plants do better in well compacted, uniform, sandy soils with relatively low porosity (high relative density) or in well-graded sands where sufficient fines (silts and clays) are present to provide moisture retention.

The converse is true for clays. High-porosity (low relative-density) clay soils allow better infiltration and water transmission to plants than do highly compacted (high relative-density) clay soils, while at the same time providing good moisture retention and storage. It is important to emphasize again the importance of compaction on soil structure and pore-size distribution in clay soils. Compaction on the wet side of optimum can reduce hydraulic conductivity (and water transmission) of a clay soil by several orders of magnitude, even when the soil is compacted to the same dry density or relative degree of compaction (see Table 3.8).

Taken as a whole, findings in the literature seem to suggest that compaction between 80 and 85 percent of the standard Proctor maximum dry density provides many of the stabilizing benefits of soil compaction without jeopardizing the viability of vegetation development and growth. Growth-limiting bulk densities or critical dry-bulk densities can readily

TABLE 3.9 Approximate Bulk Densities that Restrict Root Penetration

Texture	Critical Bulk Density for Soil Resistance gm/cm^3 (lbs/ft^3)	
	High	Low
Sandy	1.85 (116)	1.60 (100)
Coarse-loamy	1.80 (112)	1.40 (87)
Fine-loamy	1.70 (106)	1.40 (87)
Coarse-fine silty	1.80 (112)	1.30 (81)
Clayey	Depends on both clay percent and structure	

Source: From Sumner, *Handbook of Soil Science* (1999).

be compared to standard Proctor maximum dry densities. The degree of compaction suitable for root growth is calculated by dividing the critical dry density by the maximum dry density for each type of soil. Compaction rates thus calculated corresponding to growth-limiting bulk densities vary (Goldsmith et. al., 2001) from 82 to 91 percent of Standard Proctor densities, with an average of 84 percent; this limit can vary, however, depending on particular soil and site conditions.

Schor (1980, 1992) has specified and successfully revegetated fill slopes compacted to 90 percent relative compaction. His canyon fill project above the Hollywood reservoir (see Chapter 12) is a good example. Slope revegetation at this location has established and flourished on soil compacted to 90 percent of Standard Proctor. Grading specifications called for overbuilding and then scaling back the slope surface to insure achieving this degree of compaction. Part of the reason for the revegetation success lies in the shape and topography of the slopes that were designed according to landform grading principles as described in Chapter 7.

Balancing Plant Growth and Engineering Stability Needs: Limiting the density increase or degree of compaction to some predetermined threshold value is an important strategy for balancing plant growth and engineering stability requirements. However, density and/or degree of compaction are not the sole determinants of water transmission and other important soil properties. Dry-side compaction in clayey soils can result in hydraulic conductivities several orders of magnitude higher than wet-side compaction, even for two identical soils compacted to an identical dry density or relative degree of compaction (see Table 3.8). In addition, static compaction, which introduces less shear strain than kneading (or impact compaction) and results in less remolding of soil pore structure, also results in higher hydraulic conductivity and better water transmission. Accordingly, just as much attention should be paid to the manner of compaction and molding water content as to the relative degree of compaction, if the goal is to maximize plant growth as well as achieve engineering stability.

3.6.6   Management Strategies

Several different planting and/or management strategies can be employed to enhance desired characteristics of vegetation at a particular site. More vigorous and deeper rooting can be accomplished in a variety of ways, namely by:

- Watering for longer times at less frequent intervals
- Decompacting or ripping a soil before planting
- Weeding to minimize competition from unwanted plant

• Avoiding the use of overly rich topsoil dressings

Fire is often used as management tool, sometimes with unanticipated consequences. Levees are frequently fired to rid them of woody vegetation. Burning, however, promotes explosive growth of fire-adapted species that may not be the vegetation of choice for soil-erosion control.

An interesting approach to mitigate the adverse effects of vegetation on slope stability is the practice of coppicing. Coppicing is a timber harvesting or pruning method that involves the production of new trees from the old stumps (see Figure 3.12). This procedure leaves the root system intact while generating smaller, multiple stems near the cut area. Many tree species have the ability to regenerate or sprout from dormant buds along their stems and

Figure 3.12 Coppicing or cutting of tree close to the ground can produce multiple shoots, or "whips," and thereby eliminate the danger of windthrowing or toppling, preserve the benefits of roots and foliage, and also minimize view obstruction.

lend themselves to coppicing, especially northern hardwoods that have dormant buds on the lowest parts of their trunks. Examples include willows and most maples and locust trees. Some species, such as aspen, also produce new sprouts from their roots, which are referred to as root suckers.

Best results with coppicing are obtained if the stumps are cut after leaf drop in the late fall or winter. Red maples, silver maples, and black locust sprouts can grow more than 6 feet the first season. As the stump sprouts grow, they can be thinned and pruned to the desired height and number of trees per stump. Coppicing mitigates two main adverse effects from the list in Table 3.2, namely, *surcharge* (#7) and *windthrowing* (#8), whilst retaining beneficial effects. There may be some initial loss of beneficial influence *interception* (#1); but this is temporary and greatly outweighed by the attendant benefits. Coppicing allows one to retain views (a reason for tree removal on slopes), enjoy smaller trees, and retain the hydromechanical benefits provided by a tree's living root system.

Another simple yet effective management technique is to control pedestrian and vehicular traffic and in critical areas that are protected by vegetation. Coastal dunes are a good case in point. Fore dunes play a critical role in a shoreline defense system. Dune vegetation is very effective at trapping drifting sand and helping to build and accrete dunes (see Figure 3.13), but this same vegetation is very vulnerable to trampling and traffic. The use of boardwalks and walkover structures (see Figure 3.14) is an effective way of preserving the protective role of vegetation. Browsing damage to new plantings

Figure 3.13 American beach grass (dune grass) traps drifting, blowing sand and helps to build protective dunes.

Figure 3.14   Use of dune walkover to protect beach grass and other dune vegetation against foot traffic and trampling.

can be controlled by the use wire mesh cages (see Figure 3.15) or plastic sleeves placed around stems of saplings.

3.7   SUMMARY

Vegetation improves the resistance of slopes to both surficial erosion and mass wasting. Conversely, the removal of slope vegetation tends to accelerate or increase slope failures. Specific hydromechanical mechanisms can be identified by observing which vegetation affects stability in both beneficial and detrimental ways. Woody vegetation improves shallow-mass stability mainly by increasing the shear strength of the soil via root reinforcement and by a buttressing effect from well-anchored stems. Vegetation also modifies the hydrologic regime by intercepting rainfall in the foliage and by extracting and transpiring soil moisture via the roots.

The mechanical or reinforcing effect of plant roots on the stability of slopes can be described and accounted for in a systematic manner. Root fibers reinforce a soil by transfer of shear stress in the soil matrix to tensile resistance in the fiber inclusions. Simple force-equilibrium models are useful for identifying parameters that affect root reinforcement and predicting the amount of strength increase from the presence of fibers in a soil. Shear strength increases from fiber reinforcement can be incorporated into standard slope-

Figure 3.15   Wire mesh cages used to protect newly planted slope vegetation against browsing animals.

stability analyses. The presence of roots and their reinforcing effect can improve the factor of safety against sliding significantly at shallow depths.

Although woody vegetation generally has a beneficial influence on stability, it can have adverse effects as well under certain conditions. These effects include surcharge, windthrowing, and scour from large, rigid stems growing on streambanks. Other adverse or undesirable influences unrelated to stability, such as view obstruction and access hindrance, have also been cited. Many of these adverse influences can be mitigated or minimized without compromising the beneficial influence of vegetation by various strategies, including: selection of appropriate species, proper placement, grading and site preparation, optimizing compaction, and appropriate management procedures.

3.8   REFERENCES

Biddle, P. G. 1983. Patterns of soil drying and moisture deficit in the vicinity of trees on clay soils. *Geotechnique* 33(2):107–126.

Bohm, W. 1979. Methods of Studying Root Systems. In *Ecological Services* No.33. Berlin: Springer-Verlag.

Burroughs, E. R., and B. R. Thomas. 1977. Declining root strength in Douglas fir after felling as a factor in slope stability. *Research Paper INT-190*. Ogden, UT: Intermountain Forest and Range Experiment Station, U.S. Forest Service.

Brenner, R. P. 1973. A hydrologic model study of a forested and a cutover slope. *Bulletin Hydrologic Sciences* 18(26):125–143.

Coppin, N. J., and I. Richards. 1990. *Use of Vegetation in Civil Engineering.* Kent, England: Butterworths and Sevenoaks.

Daddow, R. L., and G. E. Warrington. 1983. *Growth-Limiting Soil Bulk Densities as Influenced by Soil Texture.* WDG Report, WSDG-TN-00005. Washington, DC: USDA Forest Service.

Durant, J. 2001. "What About the Soil?" *Landscape Architect and Specifier News* 17(11):28–31.

Ecabert, R. M. 1993. *Coppicing: A Management Program for Trees on Hillsides that Block Views.* Cincinnati, OH: Cincinnati Urban Landscape Tree Care Specialists.

Endo, T., and T. Tsuruta. 1969. The effect of tree roots upon the shearing strength of soil. *Annual Report of the Hokkaido Branch, Tokyo Forest Experiment Station* (18): 168–179.

Goldsmith, W., M. Silva, and C. Fischenich. 2001. *Determining Optimum Degree of Soil Compaction for Balancing Mechanical Stability and Plant Growth Capacity.* ERDC-TN-EMRRP-SR-26. Vicksburg, MS: U.S. Army Engineer Research and Development Center.

Gray, D. H. 2003. Optimizing soil compaction and other strategies. *Erosion Control* 9(6): 34–41.

Gray, D.H. and Barker, D. (2004). Root-Soil Mechanics and Interactions. In: *Riparian Vegetation and Fluvial Geomorphology: Hydraulic, Hydrologic and Geotechnical Interactions.* S.J. Bennett, J.C. Collison, and A. Simon (eds). Water Science and Application Series Volume 8, American Geophysical Union: 113-123.

Gray, D. H. and R. Sotir. 1996. *Biotechnical and Soil Bioengineering Slope Stabilization.* New York: John Wiley & Sons.

Gray, D. H., and H. Ohashi. 1983. Mechanics of fiber reinforcement in sands. *Journal of Geotechnical Engineering (ASCE)* 109(3): 335–353.

Greenway, D. R. 1987. Vegetation and slope stability. In *Slope Stability,* eds. M. F. Anderson and K. S. Richards, 187–230. New York: John Wiley & Sons.

Sumner, M. E., ed. 1999 *Handbook of Soil Science.* Boca Raton, FL: CRC Press.

Hausmann, M. R. 1990. *Engineering Principles of Ground Modification.* New York: McGraw-Hill, Inc.

Hathaway, R. L., and D. Penny. 1975. Root strength in some *Populus* and *Salix* clones. *New Zealand Journal of Botany* 13:333–344.

Henderson, R., et al. 1983. Morphology of the structural root system of Sitka spruce 1: analysis and quantitative description. *Forestry* 56(2):122–135.

Kozlowski, T. T. 1971. *Growth and Development of Trees,* vol. 2. New York: Academic Press.

Lambe, T. W. 1958. "The permeability of compacted fine-grained soils," Special Technical Publication 163, 55–67. Philadelphia, PA: American Society for Testing Materials (ASTM).

Lambe, T. W., and R. V. Whitman. 1969. *Soil Mechanics.* New York: John Wiley & Sons.

Landscape Architect and Specifier News (LANS). 2001. Structural soil. *Landscape Architect and Specifier News* 16(8):48.

Maher, M., and D. H. Gray. 1990. Static response of sands reinforced with randomly distributed fibers. *Journal of Geotechnical Engineering (ASCE)* 116(11):1661–77.

Mitchell, J. K. 1993. *Fundamentals of Soil Behavior,* 2nd ed. New York: John Wiley & Sons.

Nilaweera, N. S. 1994. Effects of tree roots on slope stability: The case of Khao Luang Mountain area, South Thailand. Dissertation No. GT-93-10. Thesis submitted in partial fulfillment of requirements for degree of Doctor of Technical Science, Asian Institute of Technology, Bangkok, Thailand.

Nolan, M. F. 1981. Vegetation on corps of engineers project levees in the Sacramento–San Joaquin Valley, California. In *California Riparian Systems Conference,* eds. R. E. Warner and K. M. Hendrix. Davis, CA: University of California, Davis.

O'Loughlin, C. L. 1974. The effects of timber removal on the stability of forest soils. *Journal Hydrology (NZ)* 13:121–134.

Patric, J. H., et al. 1965. Soil water absorption by mountain and piedmont forests. *Soil Science Society of America Proceedings* 29: 303–308.

Riestenberg, M. M., and S. Sovonick-Dunford. 1983. The role of woody vegetation on stabilizing slopes in the Cincinnati area. *Geologic Society of America Bulletin* 94:504–518.

Riestenberg, M. M. 1994. Anchoring of thin colluvium by roots of sugar maple and white ash on hillslopes in Cincinnati. U.S. Geological Survey Bulletin 2059-E. Washington, DC: U.S. Government Printing Office.

Schiechtl, H. M. 1980. Bioengineering for Land Reclamation and Conservation. Edmonton, Alberta: University of Alberta Press.

Schor, H. 1980. Landform grading: Building nature's slopes. *Pacific Coast Builder* (June):80–83.

———. 1992. Hills Like Nature Makes Them. *Urban Land* (March):40–43.

Schor, H., and Gray, D. H. 1995. Landform grading and slope evolution. *Journal of Geotechnical and Geoenvironmental Engineering (ASCE)* 121 (10):729–734.

Seed, H. B., and C. K. Chan. 1959. Structure and strength characteristics of compacted clays. *Journal of Soil Mechanics and Foundation Engineering (ASCE)* 85(5):87–128.

Shewbridge, S. E. and N. Sitar. 1989. Deformation characteristics of reinforced sand in direct shear. *Journal of Geotechnical Engineering (ASCE)* 115(GT8):1134–1147.

———. 1990. Deformation based model for reinforced sand in direct shear. *Journal of Geotechnical Engineering (ASCE)* 116(GT7):1153–1157.

Shields, F. D. and D. H. Gray. 1993. Effects of woody vegetation on the structural integrity of sandy levees. *Water Resources Bulletin* 28(5):917–931.

Sidle, R. C., A. J. Pearce, and C. L. O'Loughlin. 1985. *Land Use and Slope Stability.* American Geophysical Union, Water Resources Monograph 11. Washington, DC: American Geophysical Union, Water Resources.

Sutton, R. F. 1971. *Form and development of conifer root systems.* The Journal of Ecology, Vol. 59, No. 3, pp. 916-917, publ. by British Ecological Society, London, England .

Swanston, D. N. 1974. Slope stability problems associated with timber harvesting in mountainous regions of the western United States. *USDA Forest Service General Technical Report* PNW-21. Washington, DC.

Tschantz, B. A., and J. D. Weaver. 1988. Tree growth on earthen dams: A survey of state policy and practice. Civil Engineering Department, University of Tennessee, 36 pp. Internal publication by U. Tenn, Civ. Eng. Dept.

Tsukamoto, Y., and O. Kusuba. 1984. Vegetative influences on debris slide occurrences on steep slopes in Japan. In *Proceedings, Symposium on Effects of Forest Land Use on Erosion and Slope Stability.* Honolulu, HI: Environment Policy Institute.

Turmanina, V. I. 1965. On the strength of tree roots. *Bulletin Moscow Society Naturalists* 70 (5):36–45.

United States Department of Agriculture (USDA) Soil Conservation Service. 1978. *Predicting Rainfall Erosion Losses: A Guide to Conservation Planning.* USDA Agricultural Handbook #537. Washington, DC: USDA Soil Conservation Service.

United States Department of Agriculture (USDA). 2000. Mechanical Land Treatment. National Resources Conservation Service. Conservation Practice Specification, Code 548. Washington, DC: USDA.

Waldron, L. J. 1977. The shear resistance of root-permeated homogeneous and stratified soil. *Soil Science Society of America Proceedings* 41:843–849.

Waldron, L. J., and S. Dakessian. 1981. Soil reinforcement by roots: calculation of increased soil shear resistance from root properties. *Soil Science* 132(6):427–35.

Wu, T. H., W. P. McKinell, and D. N. Swanston. 1979. Strength of tree roots and landslides on Prince of Wales Island, Alaska. *Canadian Geotechnical Journal* 16(1): 19–33.

Wu, T. H., R. M Macomber, R. T. Erb, and P. E. Beal. 1988a. Study of soil-root interactions. *Journal of Geotechnical Engineering (ASCE)* 114(GT12):1351–1375.

Wu, T. H., P. E. Beal, and C. Lan. 1988b. In-situ shear test of soil-root systems. *Journal of Geotechnical Engineering (ASCE)* 114(GT12):1376–1394.

Ziemer, R. 1981. Roots and shallow stability of forested slopes. *International Association of Hydrological Sciences,* Publication No. 132, pp. 343-361. *Intl. Assoc. of Hydrology Scientific Press*Walllingford, Oxford, England, UK 0X10-8BB.

4 Influence of Topography on Slope Stability and Hydrology

4.1 INTRODUCTION

Perhaps no other variable affects the stability of slopes with regard to surficial erosion and mass wasting as does topography or slope morphology. Topography and hydrology also interact in a synergistic manner; the initial topography influences the development of channel networks, drainage densities, and drainage areas. These hydrologic responses in turn help determine and shape the landscape. Topographic variables include slope gradient, length, height, and profile or slope shape.

The influence of gradient on stability has been investigated at length and its effects on both surficial erosion and mass wasting quantified using various mathematical models (as described in Chapter 2). The influence of slope shape or profile, on the other hand, is not as well understood. Slope profiles considered in this regard include planar (uniform), concave, convex, and compound (convex-concave) profiles. Conceptual and mathematical models are examined in this chapter as well as the results of laboratory and field tests designed to investigate the effect of slope shape on mass stability and resistance to rainfall erosion.

The analyses considered in this chapter are two-dimensional, that is, they assume that slope profiles are invariant in all vertical planes perpendicular to one of the coordinate axes. It is very important to distinguish this type of analysis from that described in the latter part of Chapter 2, which examined stability and erosion thresholds using three-dimensional topographic elements. At first glance, this two-dimensional representation may appear to be unduly restrictive and questionable. Few natural slopes have invariant slope profiles in one direction; instead they tend to exhibit planform curvature—e.g., rounded, convex hilltops or slopes with hollows or topographically convergent areas. When groundwater is present, slopes with hollows or topographically convergent areas can lead to elevated pore pressures and slope instability, as noted in Chapter 2.

Most engineered or artificial slopes, on the other hand, are usually planar in form with an unvarying, downslope gradient and little, if any, planform curvature. Benching may be employed on occasion, but planar faces are generally the rule. These artificial slopes include embankments (dams), cut slopes along transportation corridors, cut or fill slopes within hillside developments, and earthen waste stockpiles or landfills. Topographic parameters typically considered in mass-stability analyses include slope angle and height. Topographic parameters normally considered in estimating soil-erosion losses include inclination and length of slope. Slope shape or form has seldom if ever been considered in either type of analysis in the past.

Conventional grading practice does not usually promote nor encourage other slope forms, for example, concave-, convex-, or compound-slope profiles. The reason for this grading practice and slope-form preference is somewhat puzzling. Natural slopes do not typically exhibit planar slope faces with uniform, unvarying gradients. Instead, natural slopes manifest a variety of complex slope forms and profiles. Slopes that start out with planar topography also tend to change with time into slopes with curvilinear shapes in both the downslope and cross-slope direction (Meyer and Kramer, 1969; Nash, 1977; Schor and Gray, 1996). In other words, slopes tend to evolve over time into equilibrium shapes that seldom, if ever, are entirely planar. The topic of slope evolution is discussed at length in Chapter 5.

4.2   MODELING APPROACHES AND ASSUMPTIONS

There are several ways to approach or investigate the influence of slope shape or profile. The first is simply to invoke a conceptual model, a mental image of the problem. This provides a way to think rationally about and compare the relative stability of planar versus concave slope profiles. More rigorous, theoretical analyses can also be undertaken based on physical and/or mathematical models of different slope forms. A deterministic slope-stability analysis, based on limiting equilibrium or alternatively on plasticity theory (limit analysis), are good examples of this approach. Finally, the results of field and laboratory tests of erosional losses from different slope forms can be reviewed. The findings from each of these approaches will be presented in the following sections.

An important caveat governs the analyses presented here; namely, all the slope shapes are two-dimensional. This means that a given slope profile remains invariant in all vertical planes perpendicular to one of the coordinate axes, as shown in Figure 4.1.

Changes in slope profile along the third dimension, typical of slopes in a natural watershed or catchments, can produce topographic elements that are convergent or divergent in shape. This morphological change can significantly affect both overland- and subsurface-flow patterns by elevating subsurface-pore pressures and overland-flow depths. These hydrological consequences, in turn, can affect erosion and mass stability thresholds, as explained previously in detail in Chapter 2. In general, three-dimensional convergent (i.e., "bowl"-shaped) topographical elements are more prone to slope failure and channel incision than convex or divergent elements.

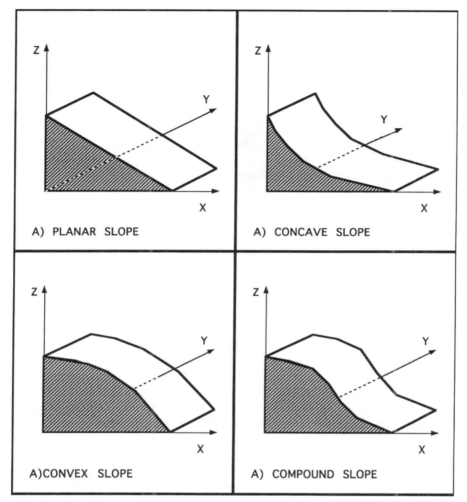

Figure 4.1 Slope shapes (planar, concave, convex, and compound) showing profile invariancy in vertical (zx) planes.

4.3 CONCEPTUAL MODELING

4.3.1 General

A conceptual model is basically a mental image of a natural phenomenon that captures the essential features or attributes of the phenomenon. In this case, we start with a schematic diagram of a simple, uniform soil slope with a planar face. The slope has a height (H) and a uniform inclination (β). For purposes of analysis, the slope is divided into a series of horizontal layers, as shown in Figure 4.2.

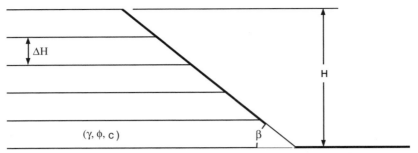

Figure 4.2 Schematic diagram of a simple slope subdivided into layers that have identical thicknesses (ΔH) and soil properties (γ, ϕ, c).

Each layer, however, has identical soil properties (γ, ϕ, c) and thicknesses (ΔH). The layering is an artificial construction that allows one to examine the influence of changing the inclination (B_i) at the face of each layer on both surficial erosion and mass wasting.

4.3.2 Mass Stability

We examine first the mass stability of the artificially layered slope shown in Figure 4.2. The lower layers are subjected to vertical (overburden) stresses from the overlying layers. The magnitude of the overburden stress is proportional to the depth, with the greatest stress acting on the lowermost layer (as shown schematically in Figure 4.3).

The mass stability of each layer will depend on the (1) soil properties of each layer, (2) layer thickness and inclination, and (3) vertical stresses on each layer. The mass stability, or safety factor against failure of each layer, can be determined by any number of classic stability analyses (Abramson et al., 2002) that takes these parameters into account. All slope and soil variables for each layer are identical; only the overburden stress differs. Clearly, the

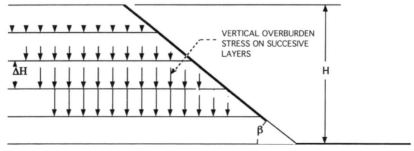

Figure 4.3 Schematic diagram of a simple slope subdivided into layers showing relative magnitude of vertical (overburden) stresses acting on each layer.

lowermost layer will have the lowest factor of safety, because it experiences the highest vertical overburden stress.

To achieve the same factor of safety for the entire slope, that is, for each layer, it will be necessary to adjust the inclination of each face. The upper layer, which is not subjected to any overburden stress, can have a relatively steep inclination; whereas the lowermost layer, which is subjected to a large overburden stress, must have a much lower inclination. This adjustment in inclination to bring the safety factor of the entire slope (all layers) to the same factor of safety is shown schematically in Figure 4.4. The stepwise adjustment in inclination now produces a slope that resembles a concave shape. In fact, subdividing the slope into layers with smaller and smaller thicknesses will ultimately produce a surface that simulates a curvilinear shape.

4.3.3 Surficial Erosion

We repeat essentially the same exercise with a layered soil slope (see Figure 4.2) that is subjected to rainfall erosion. The critical slope parameters, with regard to soil-erosion losses, are steepness and length of slope. With other factors held constant, soil-erosion losses increase with both steepness and length of slope. Increasing length means more opportunity for runoff to accumulate, resulting in larger tractive stresses acting on the slope face. Slope length can be thought of as the analogue to depth and inclination in terms of its influence on surficial erosion vis-à-vis mass stability. The relative magnitude of the tractive stresses acting on the face of each layer is shown schematically in Figure 4.5. The tractive stresses increase with increasing slope distance, down the face of the slope. Again by gradually decreasing the inclination of the faces of the layers, toward the bottom, as shown in Figure 4.6, the tractive stresses can also be decreased to a more or less constant value, thus minimizing soil loss.

In summary: A simple conceptual model demonstrates that adjusting slope angle in a downslope direction while still reaching the same toe point—i.e.,

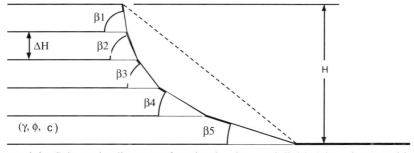

Figure 4.4 Schematic diagram of a simple slope subdivided into layers with the inclination of their faces adjusted to create the same factor of safety for each layer.

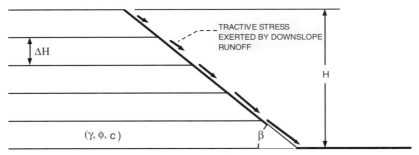

Figure 4.5 Schematic diagram of a simple slope subdivided into layers, showing relative magnitude of tractive (shear) stresses acting on face of each layer. Length of arrow denotes relative magnitude of tractive stress.

making the slope concave—will result in a more stable slope from the point of view of both mass stability and resistance to surficial erosion.

4.4   PHYSICAL-MATHEMATICAL MODELS

4.4.1   General

Surficial erosion is characterized by the movement of individual particles of soil under tractive stresses exerted by flowing water (e.g., runoff) or wind. Mass wasting (or mass movement), on the other hand, is characterized by the movement of relatively large masses of soil or rock along initially well-defined failure surfaces under the action of gravity. Mathematical models have been developed in both cases (see Chapter 2) to determine either the factor of safety against mass movement on the one hand or the likelihood of soil losses due to erosion on the other.

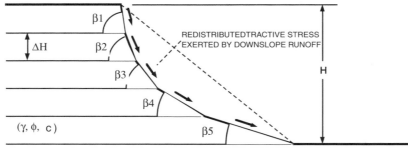

Figure 4.6 Schematic diagram of a simple slope subdivided into layers with the inclination of their faces adjusted to minimize the increase in tractive stresses in the downslope direction.

4.4.2 Mass Stability

Limiting-Equilibrium Models: A limiting-equilibrium approach is a good example of a physical-mathematical model for analyzing mass stability. Forces (or moments) resisting slope failure can be compared with forces (or moments) causing failure. as shown schematically in Figure 4.7.

The balance (or ratio) of the resisting moment (or resisting forces) versus the overturning moment (or downslope-shear forces) is the calculated factor of safety (FS) against slope failure. Various limiting equilibrium models have been developed (Abramson et al., 2002) that make different assumptions about the distribution of forces and the location and shape of the failure surface. A mathematical-model prediction is exactly correct only when the slope prototype fails and the safety factor calculated by the model is one, namely, FS = 1.0.

Soils fail when the shear stresses reach the shear strength as expressed by the Coulomb-failure criterion. There are basically two ways in which the Coulomb-failure criterion can be applied in a stability analysis, namely:

- by assuming that the failure criterion is satisfied at every point throughout a collapse zone in a soil

- by assuming that collapse occurs as a result of relative movement of rigid soil bodies and that plastic flow occurs only along the boundary between such bodies

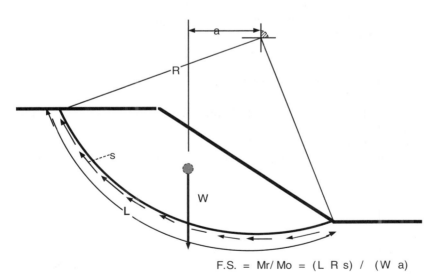

$$F.S. = Mr/Mo = (L\ R\ s) / (W\ a)$$

Figure 4.7 Mass stability of simple, cohesive-soil slope, based on balance (ratio) of resisting moments (Mr) vs. moments causing sliding (Mo) along an assumed circular failure surface.

The methods of analysis based on these key assumptions are known as the slip-line (or plasticity theory) and limiting-equilibrium methods.

Plasticity Theory: Plasticity models are more complex and less commonly used in geotechnical engineering work. As noted previously, these models assume that any infinitesimal element of soil within a collapse zone is in the Coulomb-failure state. Interior surfaces along which this condition occurs are called "slip lines." The equilibrium equation and Coulomb-failure criterion can be used to find the position and orientation of these slip lines.

A slip-line approach or plasticity solution can also be used to predict the optimal—i.e., most efficient—slope shape such that the Coulomb-failure criterion is satisfied at every point throughout the collapse zone. Most efficient in this case means mobilizing the shear strength of the soil everywhere in the failure mass (collapse zone) and not simply along a single surface, as assumed in limiting-equilibrium models.

Sokolovski (1965) has carried out such an analysis for frictional-cohesive slopes; his findings are briefly reviewed here. He first assumed that loss of limiting equilibrium would cause the slope to slip downward. He then calculated what external slope shape would be required to insure that the internal collapse-state condition would be observed within the slope. The relevant slope conditions are represented schematically in Figure 4.7.

An external, vertical-stress distribution (p) acts atop the slope. The shape of the slope face is represented by an initial angle (β_o) at the crest and a variable (obtuse) angle (β) along the length of the face as shown in Figure 4.8a. The crest angle (β_o) is uniquely related to the magnitude of vertical stress (p) at point O. Sokolovski presents a numerical solution to this problem for p = constant, angle of internal friction (ϕ) = 30 degrees. In this case, the vertical-stress distribution (p) is simply replaced by a rigid, overburden with a uniform thickness (h), as shown in Figure 4.8b.

The resulting network of characteristics (slip lines) drawn from the coordinates of the nodal points, along with the external-slope shape, is shown in Figure 4.9. In this case, the load (p) was selected such that the critical slope inclination (crest angle β_o) at point O is 90 degrees, namely,

$$h = \frac{2c \cos \phi}{\gamma(1 - \sin \phi)} \tag{4.1}$$

where h = thickness of overburden
$\quad c$ = cohesion
$\quad \gamma$ = unit weight; and
$\quad \phi$ = angle of internal friction

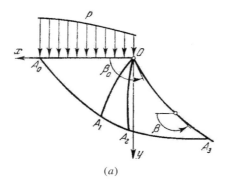

(a)

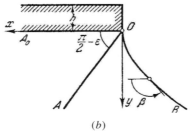

(b)

Figure 4.8 Plasticity solutions for slope subject to collapse or loss of limiting equilibrium: (a) general state of stress acting within the slope with an undetermined face shape and (b) vertical stress exerted by rigid body of uniform thickness acting atop the slope.

Numerical solutions of the same problem for different angles of internal friction (ϕ) in intervals of 5 degrees have also been carried out. The resulting shapes of the external-slope face contours are shown in Figure 4.10. The resulting, curvilinear shape of these slope contours is unmistakable. The curved, concave downward shape is particularly pronounced for small values of the internal friction angle.

In summary: Mass-stability models based on plasticity theory (limit analysis) predict that the optimal slope shape—i.e., the shape that results in complete mobilization of the shear strength throughout the collapse or failure zone—is curvilinear.

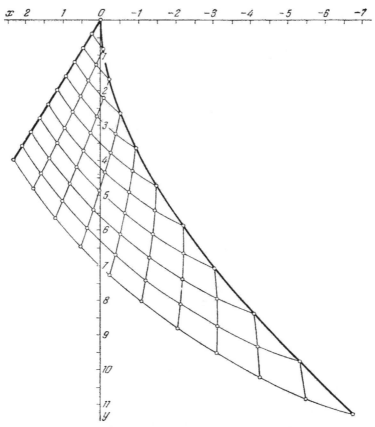

Figure 4.9   Slip lines, collapse zone, and shape of external slope contours for a soil with an angle of internal friction (ϕ) = 30 degrees and initial slope angle at crest (β_o) = 90 degrees. The coordinates (x, y) are related to actual coordinates by (x', y') by the following: $x' = (\gamma/c) x$ and $y' = (\gamma/c) y$ (from Sokolovski, 1965).

4.4.3   Surficial Erosion

Soil-loss prediction models: Soil losses caused by superficial rainfall erosion can be predicted by a semi empirical mathematical equation known as the Universal Soil Loss Equation (USLE). This soil loss equation was presented and discussed at length in Chapter 2 (see Section 2.6, Equation 2-1). All other factors equal, long slopes are more erodible than short ones (see Table 2.1). The effect of slope length on soil loss becomes slightly more pronounced the steeper the slope. The topographic (LS) factor (Israelson, 1980) can be modified relatively easily by reducing slope length—i.e., by converting a long, steep slope into a series of short, steep slopes. This is commonly accomplished in practice by

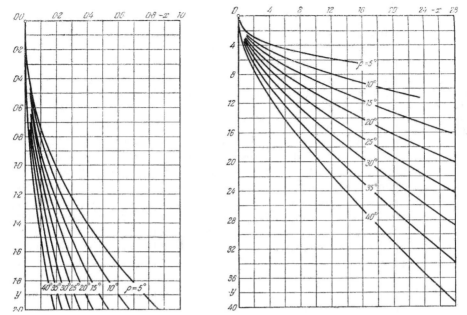

Figure 4.10   Shapes of external slope contours for soils with different angles of internal friction (from Sokolovski, 1965).

benching or terracing a slope.

The USLE does not explicitly take into account the effect of slope shape, but it does show the combined effect of slope length and angle. In this regard, it also supports the conceptual evaluation presented earlier, namely, that decreasing the gradient in the downslope direction, i.e., making the slope more concave, will tend to counteract the influence of increasing slope length.

Additional support for superior erosion resistance of curvilinear (concave) slopes comes from computer modeling of slope evolution (Nash, 1976; Meyer and Kramer, 1969). Nash (1976) showed that transport limited, planar slopes evolve over time into concave shapes along their bottom halves. Lower soil-erosion losses on concave slopes also conform with the results of computer modeling studies by Meyer and Kramer (1965). Both these studies indicate initial shape of a hillside affects (1) the rate of erosion at different locations along a slope and, ultimately, (2) how the hillside slope changes its shape as erosion progresses.

Meyer and Kramer (1969) computed soil losses from four slope shapes—uniform (planar), concave, convex, and complex (upper half concave, lower half convex). The slopes were studied at mean slope steepnesses of 10 percent. All slopes had a 20-foot elevation difference between top and bottom as shown in Figure 4.11. Soil losses were computed using soil-loss prediction equations similar to those used to generate Table 2.1 in Chapter 2.

Elevations at 10-foot intervals along each slope were used by their computer program to determine the total erosion or sediment load at the end of each 10-foot increment, based on the steepness at (and the length from the top of slope) to that point. The net erosion (or deposition) between successive increments was the difference in the sediment load at these points. Erosion depth at each location was determined as the sum of the net erosion for the two adjacent increments divided by their total lengths. Slope profiles of the different shapes developed differently, because erosion depth along each profile varied considerably. Sediment loads generated during the first erosion period along the four shapes at 5 percent average steepness are shown in Figure 4.12. The relative depth of erosion (or deposition) is indicated by the difference between the load at successive locations. Where the slope of the curve is positive, erosion is occurring; where negative, deposition is occurring. New elevations and degree of steepness were computed at all points for each successive erosion period. Their computer program also produced the input data for an electronic plotter to graph sediment loads and resulting

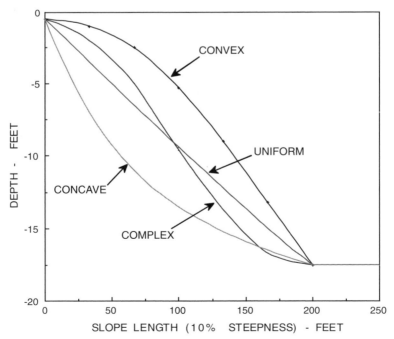

Figure 4.11 Four slope shapes at 10-percent average steepness used in computer modeling study. Vertical scale is expanded. A flat, level area was assumed beyond the toe of the slopes (from Meyer and Kamer, 1969).

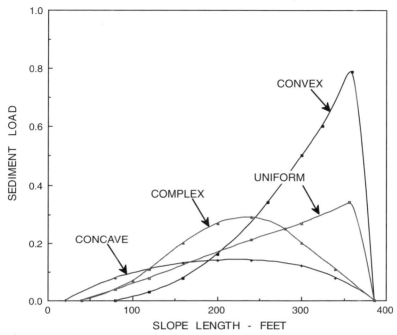

Figure 4.12 Predicted sediment load or total erosion (tons per foot of slope width) along the four original slope shapes at 5 percent steepness during the first erosion period. The sediment load increased (and net erosion occurred) where curves have positive slopes. Deposition occurred where curves have negative slopes (from Meyer and Kamer, 1969).

slope profiles.

Sediment load was low at the upper end of the convex slope where the steepness was low, but it increased rapidly as slope steepness and slope length (a surrogate for quantity of runoff) increased. In contrast, the concave slope was steepest where the least runoff occurred, namely, at the upper end of the slope. The steepness then decreased as the length of slope (runoff) increased so that sediment load at any point was low on the concave slope.

The total soil loss (indicated by the area under each curve) was much lower for the concave than for the convex and uniform slopes. Accordingly, the computer modeling predicted not only a much lower total sediment load (soil loss) for the concave slope but also a slope shape that changed less rapidly or was more stable over time than the uniform or convex shape.

Digital Terrain Model Predictions: By using a computer-generated digital terrain model, Hancock et. al. (2003) demonstrated that concave slopes exhibit greatly reduced erosion rates on rehabilitated (postmining) landforms and that concave-hillslope profiles blended in much better with the surrounding natural terrain. They noted, furthermore, that soil-mantled, fluvial erosion–dominated catchments generally have a convex upper hillslope profile with a concave profile downslope, as shown in Figure 4.13.

This profile can be described by the "area-slope" relationship, which is a widely recognized geomorphic relationship between the area draining through a point on the slope versus the slope at the point. The relationship has the form:

$$A^{\alpha}S = \text{constant} \tag{4.2}$$

where A = contributing drainage area to point of interest;
      S = slope (gradient) of point of interest;
      α = concavity of hillslopes in a natural catchment (values range from 0.2 to 0.7)

Hancock et al. (2003) used a digital terrain model known as SIBERIA to predict sediment loss from their constructed landforms. SIBERIA is a physically based predictive model that can simulate the geomorphic evolution of landforms subjected to fluvial erosion and mass transport processes (Willgoose et al., 1991). SIBERIA links generally accepted hydrologic-response models under the action of runoff and erosion over variable time scales. This means that a three-dimensional landform can adjust itself over time to erosion and deposition occurring at different locations. Consequently, SIBERIA allows the user to see how a land-form evolves or changes spatially over time. This topic is discussed in greater detail in Chapter 5, which deals with landscape evolution and equilibrium landforms. SIBERIA also permits the user to construct slopes with different initial profiles, say concave versus planar, and ex-amine how these evolve over time and how much sediment or soil loss each generates.

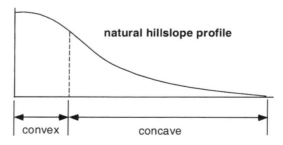

Figure 4.13 Idealized cross section of a natural hillslope in a soil-mantled landscape (after Hancock et al., 2003).

Sediment losses from linear (planar) and concave slopes in one of their simulations are compared in Table 4.1. The concave profiles resulted in greatly reduced sediment losses from their simulated hillslope . . . in some cases achieving an average five-fold reduction as shown in Table 4.1. Increasing the slope concavity while maintaining the slope length constant ($L = 200$ m) also decreased the soil loss as shown in Table 4.2.

4.5   LABORATORY AND FIELD TESTS

4.5.1   General

Conceptual and mathematical models are more convincing if they are validated by the results of laboratory and field tests. It is instructive, therefore, to examine the findings from such studies.

4.5.2   Mass Stability

With the exception of plasticity solutions, the effect of slope shape has not been considered in physical and mathematical stability analyses. As a result, there has been little impetus to establish field trials to test the effects of curvilinear slope contours on mass stability. Test slopes with different slope shapes, but identical in every other respect, could be established for this purpose. Failure would be triggered by simply loading the top of the slope with an increasing, uniform vertical-stress distribution. The slope shape requiring the highest surcharge to cause failure might then qualify as the safest or most secure slope. Such a study has yet to be carried out or, at least, yet to be reported in the technical literature.

4.5.3   Surficial Erosion

Unlike mass stability, both laboratory and field studies have been conducted to determine the influence of slope shape on surficial erosion losses caused

TABLE 4.1   Sediment Loss from Linear vs. Concave Slopes

Average Slope (%)	Slope Length (m)	Sediment Loss (t/ha/yr)	
		Linear Slope	Concave Slope
20	200	22	4
25	170	34	6
35	120	69	12
45	90	100	21

Source: From Hancock (2003).

TABLE 4.2   Sediment Loss for Different Concave Slopes

Slope Concavity α	Sediment Loss tons/hectare/year
200	4
170	6
120	12
90	21

Note: Alpha (α) is the exponent in the area-slope relationship.
Ave. slope of 20 percent over a 200 m hill slope length

Source: From Hancock (2003).

by rainfall. Shaefer et al. (1979) evaluated the effect of both slope shape and drainage network densities on erosion losses from mine-reclamation sites. They noted that slope geometry could be an important aspect of controlling erosion on mined lands. Hydrologic studies consistently indicated that a "graded stream," having a concave profile, is the most efficient for runoff transport with minimum erosion. The same held true for hill- and ridge-side slopes.

Young and Mutchler (1969) measured soil loss and runoff from concave, uniform (planar) and convex slopes in the field subjected to simulated rainfall. Their study site was located in South Dakota on deep loess soil. The uniformity of the soil to a depth of at least 5 feet allowed mechanical shaping of concave, uniform, and convex slope shapes. Transport of eroded particles occurred primarily by raindrop splash at the top and then via rills over the remainder of the slope. For slopes of equal average steepness, a concave shape greatly reduced the total amount of sediment lost from the slope in comparison with that from either a uniform or a convex slope.

Micro relief or elevation changes in the eroded soil plots are shown in Figure 4.14. Surface-elevation decline on the uniform plots was less than ½ inch at the top and increased to over 1 inch at about 62 feet. This finding reflects the effect of increasing slope length, which allows for greater accumulation or depth of runoff and which, in turn, promotes further erosion.

The decline in surface elevation of the upper ends of the concave plots was more than twice that at the upper ends of the convex plots. This finding reflects the influence of slope steepness at the top. On the other hand, the surface elevation decline (erosion) became much less farther downslope on the concave slope, and deposition actually began about 13 feet from the bottom. This occurred because of a drop in velocity and transporting capacity in the rills, as the slope leveled out toward the bottom.

D'Souza and Morgan (1976) describe laboratory experiments to determine the relationship between soil loss and slope steepness on slopes both curved and straight in profile. They included the effect of slope gradient as well and measured soil loss under simulated rainfall on thirteen different slopes: five

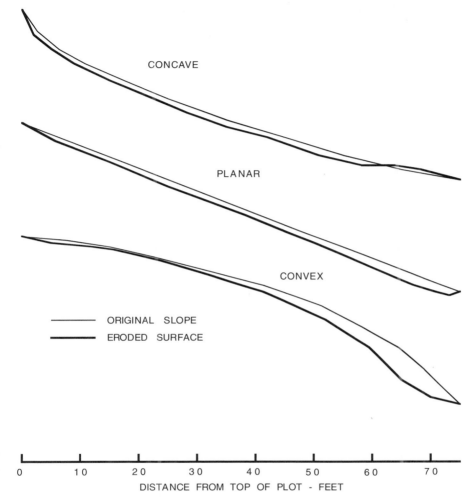

Figure 4.14 Schematic diagrams of elevation changes of slopes with concave, planar, and convex profiles (adapted from Young and Mutchler, 1969).

straight slopes at 3.5, 7, 10.5, 14, and 18 percent and four convex and four concave slopes of average steepness of 3.5, 7, 10.5, and 14 percent. On the curved slopes, the curvatures or rates of profile change were 1, 2, 3, and 4 degrees per meter, respectively. Only slope shape, steepness, and curvature were varied. Conditions of soil type, soil compaction, moisture content, rainfall intensity, and slope length were constant for each treatment.

The values of soil loss for each treatment are summarized in Table 4.3. Total soil losses on convex and straight slopes for the two less steep slopes (3.5% and 7%) were similar and much greater than those on concave slopes.

TABLE 4.3   Total Soil Losses from Slopes with Different Shapes and Degrees
of Steepness

Slope (%)	Slope Shape (Soil Loss in Grams)		
	Convex	Straight	Concave
18	—	1373	—
14	1347	864	636
10.5	956	717	530
7	678	683	518
3.5	617	622	502

Source: From D'Souza and Morgan, 1976.

On the steeper slopes, for each slope percentage, the loss was greatest on convex slopes and decreased on passing, first, to straight and then to concave shapes. Soil losses were about 20 to 25 percent higher on the straight as compared to the concave slopes over the range of slope inclinations tested in the laboratory.

Calculated depths of flows (and therefore tractive stresses) tended to increase downslope on all slopes. This finding underscores the role of slope length, and longer length allows for accumulation and greater depth of runoff. Runoff velocities increased downslope on the convex and straight slopes, and the rate of increase was greater on the convex slopes. In contrast, velocities gradually decreased downslope on the concave slopes. Deposition at the toe of concave-slope shapes was also noted in the laboratory studies. This depositional location conforms with the earlier field experiments (see Figure 4.14) conducted by Young and Mutler (1969), and it is related to the reduction in velocity and transporting capacity of the runoff on the gentler gradient.

4.5.4   Equilibrium Profiles of Natural Slopes

Visual observations of natural slopes provide another way of gauging the long-term stability of different slope profiles or shapes. A good place to observe these profiles is in steep, mountainous terrain or arid desert climates, where the absence or scarcity of significant vegetal cover make it easier to determine slope forms. Equilibrium-slope forms can be observed readily in the mesa and canyon country of the southwestern United States. A topmost, resistant layer of hard sandstone or igneous rock typically results in erosional remnants—mesas and buttes with near vertical rim walls at the top and foot slopes below that characteristically develop a concave, equilibrium profile over time, as illustrated in Figures 4.15 and 4.16.

Concave foot slopes are well developed and can be readily observed in the Unaweep Canyon a few miles south of Grand Junction, Colorado. This canyon cuts through the Umcompahgre Plateau. At one time, the ancestral Gunnison River flowed through the Unaweep Canyon; but the rate of uplift in the sur-

Figure 4.15 Photo of Rocky Mountains, Alberta, Canada, showing foot slope with concave-slope profile.

Figure 4.16 Photo of mesa with vertical rim wall at top and concave-foot slope below, Highway 128, Colorado River Canyon, Colorado.

rounding plateau was too high to allow the river to maintain its course through the canyon. As a result, the river was forced to change course and flow elsewhere (Lohman, 1981). The present canyon or valley has a saddle or high point with a relatively small stream draining away in opposite directions. These conditions have eliminated stream erosion and downcutting as a major geomorphic force in the canyon in recent geologic time. Instead, more diffuse erosional slope processes and mass wasting dominate topographic development. Over geologic time these slope processes have produced a relatively broad valley with concave foot slopes along canyon margins, as illustrated in Figure 4.17. The geomorphic evolution of slopes is treated in greater detail in Chapter 5.

4.5.5 Summary

The results of laboratory and field studies clearly show that concave slope profiles appear to be more stable and generate less sediment than uniform, planar slopes. These findings are consistent with the conceptual models discussed that explored the effect of slope shape, and they also conform with results of computer modeling of soil erosion on slopes with irregular shapes (Meyer and Kramer, 1965) and time-evolved digital terrain models (Hancock et. al., 2003). There do not appear to be any reported laboratory or field trials that have examined the effect of slope shape on mass stability; however,

Figure 4.17 Photo of Unaweep Canyon showing development of broad valley and concave foot slopes, Highway 141, Colorado.

relatively simple field trials could be established for this purpose. On the other hand, both conceptual and mathematical models (Sokolovski, 1965) indicate that concave slope profiles are superior with respect to mass stability. It should be noted that these findings are restricted to two-dimensional slopes, that is, slopes with invariant slope profiles along one of the coordinate axes.

4.6 ROLE OF DRAINAGE NETWORKS AND DRAINAGE DENSITIES

As noted at the beginning of this chapter, topography and hydrology interact synergistically to shape the landscape. This interaction is particularly signif-icant in the case of manufactured slopes and post-strip-mining reclamation landforms. Natural watersheds tend to have drainage networks and slopes that are relatively stable with respect to one another. Forces associated with rates of runoff and hillside gradient or slope, which cause channels (gullies) to develop, tend to be in dynamic equilibrium with forces that resist erosion, such as shear strength or allowable tractive stress. Under these conditions, deep incision or channel erosion is minimized, and sediment yield originates primarily from rainfall erosion processes, such as raindrop splash, sheet, and rill erosion.

Studies of postmining-reclamation landforms (Schaefer et al., 1979) often reveal a different picture. Inspection of reclamation sites in the Midwest showed, for example, that even when graded to acceptable specifications, freshly graded surfaces may be left with a high potential for advanced erosion, including channel incision and gullying. These erosional processes, in turn, can disrupt attempts at revegetation and produce sediment that must be con-tained in sediment ponds or behind check dams to prevent off-site damage. These postmining-reclamation sites were characterized by drainage-network and slope-profile characteristics that differed substantially from those of sta-ble, natural watersheds or land surfaces.

4.6.1 Drainage Density and Zero-Order Watershed

Two important hydrologic parameters that characterize drainage networks and erosional tendencies in a watershed are: (1) drainage density and (2) zero-order watershed. Both of these parameters can be measured and/or calculated in the field for both natural watersheds and constructed (reclaimed) land-scapes. Major differences in these hydrologic parameters between natural ver-sus constructed land surfaces can explain why advanced erosion occurs in one case but not another.

4.6.1.1 Drainage Density: A drainage network with sufficient capacity to carry runoff from the area must ultimately develop on any constructed or altered land surface. The relative magnitude of sediment volume transported by this network is related to the drainage density (D_d) that is defined as:

$$D_d = \Sigma \ L/A \qquad (4.3)$$

where $\Sigma \ L$ = summation of all stream segments (miles) in the drainage area, and

A = drainage area (square miles)

Natural drainage densities for areas throughout the United States are listed in Table 4.4. Drainage densities diverge widely, ranging from a low of 1.5 miles per square mile in the glaciated terrain of northern Wisconsin to a high of 650 miles per square mile in the Badlands of Perth Amboy, New Jersey.

In the absence of an adequate channel network, as expressed via drainage density, a potential source of sediment can be eroded in the process of cutting channels to convey runoff. Schaefer et al (1979) have shown that channel development is in fact a major source of sediment supply for areas without an adequate drainage network. Accordingly, a drainage network with the requisite drainage density must be designed and placed during building of constructed landforms and associated surfaces in order to reduce sediment sources.

4.6.1.2 Zero-Order Watershed: The zero-order watershed is the minimum drainage area (A_0) from which the runoff produced has sufficient force to initiate channel development (see Figure 4.18). This means that a zero-order watershed is free of channels (or gullies) and that the smaller this subwatershed becomes the greater the amount of channel incision and development that will occur in higher order watersheds which contain these zero order, subwatersheds.

The drainage density (D_r) can be related to the zero-order drainage area using the law of basin areas (Schaefer et al., 1979) as follows:

TABLE 4.4   Natural Drainage Density Values

Location	D_D miles/sq. mile
Ozark Plateau, Missouri	17
Gulf Coast Plain, Louisiana	5
Coastal Plain, Virginia	6
Appalachian Plateau, Kentucky	56
Ozark Plateau, Illinois	15
Verdugo Hills, California	26
Badlands, Perth Amboy, New Jersey	650
Badlands, Petrified Forest, Arizona	250
Coastal Range, California	35
Cumberland Plateau, Tennessee	5

From Schaefer et al., 1979.

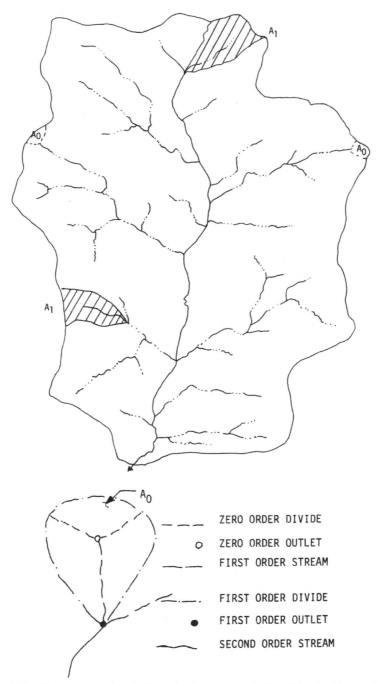

Figure 4.18 Zero-order watershed concept showing relationship of various watershed divides, outlets, and streams.

$$A_u = A_0 R_a^u \qquad (4.4)$$

where A_u = drainage area of watershed of order u (square mile) and
$\qquad R_a$ = geometric ratio

Schaefer et al (1979) point out that the drainage density (D_r) generally remains relatively constant throughout a watershed (for basins of order one and greater). Furthermore, the drainage density for the first order watershed (of area A_1) is basically the same as that for the entire watershed. Substituting for the drainage density of the first order watersheds using Equation (4.3) yields the following:

$$D_d = \frac{C_1 N_1}{A_1 N_1} \frac{C_1}{A_0 R_a} \qquad (4.5)$$

where C_1 = average first order stream length (mile)
$\qquad N_1$ = number of first order stream lengths and basins

Equation (4.5) demonstrates that the drainage density of a watershed (D_d) is inversely proportional to the first order watershed area (A_0). This means that when A_0 is small, there will be greater channel development, that is, first-order streams will occur more often and will intersect to form second-order streams more quickly. The overall effect is that a denser network will result because this process will be repeated with the higher order streams.

4.6.2 Influence of Watershed Parameters

Schaefer et al (1979) also derived an equation relating the size (area) of the zero order watershed (A_0) to physical parameters that both produce and resist erosion in channels. Mathematically, this equation is:

$$\frac{1}{A_0} = \frac{2.75 \times 10^6 \, n \, \sigma}{C_1} [S_h^{1.67} / \tau_c^{2.17}] \qquad (4.6)$$

where A_0 = zero order drainage area (square miles)
$\qquad n$ = Manning's roughness coefficient at the outlet
$\qquad \tau_c$ = critical tractive stress (pounds/square feet)
$\qquad \sigma$ = dominant runoff rate approximated by the 2-year frequency run-off event (inch/hour)
$\qquad S_h$ = average hillside slope or gradient (feet/feet).

Substitution of Equation (4.5) into Equation (4.6) yields an expression relating the drainage density (D_d) to physical elements that control channel erosion, as follows:

TABLE 4.5 Relationship of Zero-Order Watershed Area and Drainage Density to Watershed Parameters

Parameter Trend	A_0 Trend	Drainage Density
τ_c decreases	decreases	increases
S_h increases	decreases	increases
σ increases	decreases	decreases

From Schaefer et al., 1979.

$$D_d = \frac{2.75 \times 10^6\, n\, \sigma}{R_a\, C_1}\, C_1[S_h^{1.67}/\tau_c^{2.17}] \qquad (4.7)$$

Based on Equations (4.5) and (4.7), the relationship between zero-order watershed area (A_0)and drainage density (D_d) to watershed parameters (τ_c, S_h, and σ) can be summarized in Table 4.5.

4.7 REFERENCES

Abramson, L. W., T. S. Lee, S. Sharma, and G. M. Boyce. 2002. *Slope Stability and Stabilization Methods.* 2nd ed. New York: John Wiley & Sons.

Dietrich, W. E., C. J. Wilson, D. R. Montgomery, J. McKean, and R. Bauer. 1992. Erosion thresholds and land surface morphology. *Geology* 20: 675–679.

Dietrich, W. E., C. J. Wilson, D. R. Montgomery, and J. McKean. 1993. Analysis of erosion thresholds, channel networks, and landscape morphology using a digital terrain model. *The Journal of Geology* 101:259–278.

D'Souza V. P., and R. P. Morgan. 1976. A laboratory study on the effect of slope steepness and curvature on soil erosion. *Journal of Engineering Research* 21:21–31.

Hancock, G. R., R. Loch, and G. R. Willgoose. 2003. The design of post-mining landscapes using geomorphic guidelines. *Earth Surface Processes and Landforms* 28:1097–1110.

Israelson, E. 1980. Erosion control during highway construction—Manual on principles and practices. Transportation Research Board, National Co-operative Highway Research Program Report 221, TRB, Washington, DC.

Lohman, S. W. 1981. Ancient drainage changes in and south of Unaweep Canyon, southwestern Colorado, eds. R. C. Epis and J. F. Callender, 137–143. Western Slope Colorado: New Mexico Geological Society Guidebook (32nd Field Conference).

Meyer, L. D., and L. A. Kramer. 1969. Erosion equations predict land slope development. *Agricultural Engineering* 50:522–23.

Montgomery, D. R., and W. E. Dietrich. 1994. A physically based model for the topographic control on shallow landsliding. *Water Resources Research* 30(4):1153–1171.

Nash, D. B. 1977. The evolution of abandoned, wave-cut bluffs in Emmet County, Michigan. PhD dissertation, Univ. of Michigan, Ann Arbor, Michigan.

Schaefer, M., D. Elifrits, and D. J. Barr. 1979. Sculpturing reclaimed land to decrease erosion. Symposium on Surface Mining Hydrology, Sedimentology and Reclamation, University of Kentucky, Lexington, Kentucky, December 4–7.

Schor, H. J. 1992. Hills like nature makes them. *Urban Land* (March):40–43.

Schor, H. J., and D. H. Gray. 1995. Landform grading and slope evolution. *Journal of Geotechnical Engineering (ASCE)* 121(GT10):729–734.

Sokolovski, V. V. 1965. *Statics of Granular Media.* New York: Pergamon Press.

Willgoose, G. R., R. I. Bras, and I. Rodriguez-Iturbe. 1991. A physically based coupled network growth and hillslope evolution model: 2. Applications. *Water Resources Research* 27(7):1685–1696.

Young, R. A., and C. K. Mutchler. 1969a. Effects of slope shape on erosion and runoff. *Transactions of the American Society of. Agricultural Engineers* 12:231.

Young, R. A., and C. K. Mutchler. 1969b. Soil movement on irregular slopes. *Water Resources Research* 5(5):1084–89.

5 Geomorphic Evolution of Slopes

5.1 INTRODUCTION

Landscapes are composed of assemblages of landforms, and landforms tend to be bounded or defined by slopes. A good understanding of landscape evolution, therefore, requires knowledge of how slopes evolve and what processes control their development. This same knowledge and understanding are required for the design of artificial landforms, where sediment reduction and long-term slope stability are important goals. Sharp (1984) provides a review of the role that slopes play generally in landscape evolution and how climate, lithology, and vegetation affect their development.

One of the most important aspects of slopes is that they are seldom static, at least in the geological sense. They may appear to attain a quasi-steady form; but, in fact, they undergo subtle changes over time as a result of a number of geological processes that are constantly at work. This appearance of steady-state form may account for the fact that conventional engineering stability analyses do not take time into account. Neglect of slow, time-dependent geological processes on slope morphology, such as weathering and creep, may be acceptable in the short term, say less than 10 years, but could lead to serious consequences in the long term, say 100 years or more. The latter time frame is applicable to large earth dams and to constructed landforms containing toxic ore tailings or hazardous wastes.

Considerable controversy has surrounded the geomorphic development and evolution of slopes. One school of thought promulgated at the turn of the twentieth century by its chief exponent W. M. Davis (1898) basically claims that slopes evolve by a general process of "downwearing." Under this scheme of landscape evolution, an uplifted land mass of gentle relief is dissected by streams into a complex of deep, narrow canyons and sharp ridges. According to Davisian concepts, this landscape is then worn down by weathering and erosion, with the slopes becoming progressively less steep. The end product is a new and lower surface of gentle relief graded to the ultimate base level of erosion, namely, the sea. Another school of thought promulgated a half century later, by its chief exponent Walter Penck (1953), postulated that slopes evolved by a general process of "backwearing." According to this concept, plainlike erosion surfaces are formed not by downwearing but rather by backwearing as a result of the retreat of steeper slopes that maintain a relatively constant form as they recede, that is, the slope face retreats in a parallel fashion.

119

In his review of landscape evolution, Sharp (1984) maintains that the conflict between the proponents of "backwearing" and "downwearing has been grossly overemphasized and that evidence for both modes can be found in natural settings. He cites examples where each mode appears to occur in both arid and humid environments but with a difference in emphasis.

5.2   ROLE OF GEOLOGIC PROCESSES

The earth's surface features or landscape are affected by geologic processes that operate over different time scales and at different locations. These processes determine the shape, structure, and composition of the earth's crust. Geologic processes can be classified into one of two major types—i.e., *endogenetic* and *exogenetic*.

1. *Endogenetic: Internal earth movement*
   a. Characteristics/Significance
      • Operate and/or originate deep within the earth's mantle and responsible for creating major planetary landforms, e.g., mountain ranges, plateaus, and basins. Difficult to predict and largely beyond the control of humans.
   b. Examples
      • Seismicity (earthquakes)
      • Vulcanism
      • Diastrophism (folding and faulting, tectonic activity)
2. *Exogenetic: Surface processes*
   a. Characteristics/Significance
      • Derive energy from climate and endogenetic processes, erode and transport earthen materials, sculpt and mold surficial landforms. These are the primary processes of concern to engineers and geomorphologists, because they can be controlled and modified to some extent.
   b. Examples

Process	Typical landform
• fluvial	river terrace, delta
• lacustrine	beach ridges
• aeolian	sand dune
• glacial	moraine
• glacio-fluvial	esker, kame
• gravitational	talus slope, colluvial hillslope hollow

5.3   GEOMORPHOLOGY

Geomorphology is a branch of geology that deals primarily with the surface of the earth and geologic processes that operate on and shape the earth's surface. Geomorphology is basically a study of landforms and landscape evolution. Specific landforms result from the complex interactions between exogenetic geologic processes and various types of earthen materials (rocks and soil) and vegetation. Hancock et al. (2003) have described the process of landform evolution simply and succinctly in the following way:

> The undisturbed landscape is an expression of the integrated effect of climate, geology and vegetation over geological time.

This integration or functional dependence can be expressed by a very generalized mathematical expression (Fookes and Vaughan, 1986), such as:

$$F(t) = \int (PM)\, dt$$

where $F(t)$ = time dependent landform
M = material type
P = geologic process (exogenetic)

This equation captures the essential nature of the interaction and its time dependency. On the other hand, the relationship is not very useful because it lacks specificity and predictive qualities. It would be helpful to know, for example, how an artificial landform, say a man-made tailings embankment, will evolve over time. Or what type of landform design might be superior in terms of its resistance to erosion, landsliding, and incision. This predictive capability has been developed only recently (Evans et al., 2000; Hancock et. al., 2003) with the advent of digital terrain models that make it possible to link changes in spatial coordinates with geologic process equations.

5.4   SLOPE ATTRIBUTES AND CHARACTERISTICS

5.4.1   Classification of Slopes

Slopes vary in size, shape, pattern, and declivity, the last ranging from vertical to almost flat. Slopes can be classified on the basis of a number of attributes, for example, whether they are (1) primary or secondary, (2) bedrock or detrital, (3) wash versus gravity, and (4) supply, transport, or accumulation slopes. A brief discussion of each of these classification terms follows:

Primary vs. secondary slopes: Some slopes are created in their present form principally by endogenetic geological processes, e.g., faulting, folding, or vulcanism. Shapes of primary slopes are modified relatively quickly by endogenetic processes, e.g., surficial erosion and mass movement, so they rapidly become secondary in form.

Bedrock vs. detrital slopes: Many slopes are characterized by a combination of bedrock outcrops and regolith (soil) accumulations. It is sometimes useful to speak of "bedrock slopes" as opposed to "detrital slopes" in those cases where one or the other dominates.

Wash vs. gravity slopes: Surficial erosion often plays a major role in shaping slopes. Accordingly, it is reasonable to refer to "wash slopes" (those slopes or portions thereof on which the work of surface water predominates) as opposed to "gravity slopes," which are shaped primarily by the downhill creep of loosened material.

Supply, transport, or accumulation slopes: Supply slopes are portions of the slope from which debris is derived; whereas accumulation slopes are portions where material is deposited. Transport slopes, on the other hand, are portions of a slope across which debris and sediment move. The declivity and dominant geologic process acting on each of these slope elements also varies. With regard to declivity, supply slopes tend to be steep; whereas accumulation slopes are relatively gentle and usually decrease in declivity in a downslope direction. Transport slopes tend to be nearly uniform in gradient and intermediate in declivity. Supply slopes are commonly bare rock or dominated by erosion so that little if any sediment remains there. Transport slopes are mostly mantled by a thin but continuous mantle of loose material in transit. Accumulation slopes typically are underlain by thicker deposits of sediment that undergoes some slope washing.

5.4.2 Slope Profiles and Elements

The main morphometric attributes of slopes are their declivity and shape (or profile). Most engineered slopes (and relatively few natural ones) have rectilinear profiles with uniform declivities. Most natural slopes, on the other hand, have some curvature, either convex or concave to the sky. Convex slopes are usually dominated by erosion, rectilinear slopes are primarily transportive, and concave slopes can be either erosional or depositional, frequently the latter (Sharp, 1984). The combination of convex, rectilinear, and concave elements into a smoothly integrated, compound slope profile (see Figure 5.1) is a very common, and relatively stable shape in natural landscapes.

Slopes can be broken down into morphologic units or elements (Haigh, 1978), and each unit can undergo specific changes in location and/or shape over time, depending on whether degradation or accretion occurs.

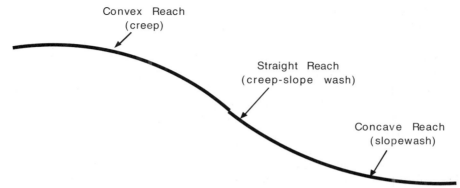

Figure 5.1 Compound slope profile on which creep dominates the convex reach, surface wash dominates the concave reach, and creep and wash are roughly balanced in the straight reach (after Sharp, 1984).

1. *Rectilinear units* (slope segments)
   . . . are subject to parallel advance or retreat or to ($+$) or ($-$) slope declines
2. *Curvilinear units* (slope elements)
   . . . are subject to ($+$) or ($-$) variations in radius of curvature

5.4.3 Slope Processes

Slopes degrade or wear away over time by the combined geologic processes of weathering and transportation. Weathering, both physical and chemical, disintegrates and decomposes bedrock so that the weathering products can be removed and transported downslope. Chemical weathering, which requires the presence of water, predominates in most regions. Transportation begins once particles are loosened or released from a consolidated substrate by weathering.

Most downhill transport on slopes results from two major geologic processes—surficial erosion and mass movement. Surficial rainfall erosion occurs in many different forms, including raindrop splash, sheet flow, rilling, and gullying. These types of rainfall erosion and their controlling factors are reviewed in detail in Chapter 2. Most erosion on slopes is believed (Sharp, 1984) to result from two processes, that is, surface wash accomplished by water delivered to a slope largely by rainfall and creep accomplished by the slow downslope movement of soil by gravity. Creep is a form of mass movement; other forms include debris slides and earth flows. The latter occur relatively rapidly and are dominant only locally or under special circumstances. Various types of mass movements and their characteristics and controlling factors are described in Chapter 2.

The literature on slope evolution often uses the terms "diffusive" and "fluvial," respectively to describe geologic processes that act upon and degrade slopes. Terrestrial landscapes and landforms consisting of hills, mounds, embankments, and upland slopes (including valley sides) are acted upon primarily by diffusive and fluvial processes respectively. Diffusive processes include slope wash and creep. Slope wash refers to rainfall delivered to a slope that impacts on and runs off, carrying soil particles with it. Slope wash includes raindrop splash, sheet flow, and rilling. Fluvial processes, on the other hand, are characterized by pronounced incision and formation of channels, e.g., gullying and stream-channel erosion. Rills result when fine threads of water coalesce into streamlets that form small, ephemeral channels. Rills are a form of surficial erosion that represents a transition from diffusive to fluvial erosion. Biogenic activity (e.g., animal burrowing) can be considered as a diffusive process in terms of its impact on slopes.

Diffusive processes tend to degrade (smooth, round, and lower) slopes over time whereas fluvial processes incise the landscape, resulting in the formation of channel networks that convey both sediment and water. Wilgoose (1994) comments on the implications of these competing processes as follows:

> Long-term landscape evolution is the balance between fluvial processes that incise the landscape and diffusive processes that . . . round or smooth the landscape.

Montgomery and Dietrich (1989) describe this balance in a slightly different manner:

> Channel head locations are controlled by the competing influences of fluvial transport in channels and diffusional hillslope transport processes which concentrate colluvium in hollows and tend to infill channels.

5.5   APPROACHES TO SLOPE EVOLUTION PREDICTION

Four basic approaches for studying and predicting slope evolution have been identified by Haigh (1978). These approaches, their advantages and limitations, are cited below.

5.5.1   Traditional Approach

Deductive and mathematical modeling of slope profiles. These profiles can be expressed by mathematical (or geometric) equations based on a variety of assumptions. This approach oversimplifies complexity of slope evolution and overlooks factors that affect change in slope geometry over time.

5.5.2   Morphometric Approach

Mathematical modeling of slope profiles that takes time into account. This approach attempts to classify geometric properties of slope profiles and chart their evolution in terms of subsequent geometric alterations, for example, modeling using the diffusion equation (Nash 1977). This appears more fruitful than the traditional approach but still does not explain the interaction between slope processes and materials.

5.5.3   Process Approach

Analysis of the role of individual processes and control components governing erosion and study of interactions and their relative importance. Significant laboratory and field studies have been carried out, but slope evolution is governed by complicated processes and the relationship between these processes and various control components are not yet completely understood. Parsons (1988) has pointed out other limitations to process-response models, namely, the effect of changes in climate over time and changes in conditions at the base of slopes, e.g. removal of landslide debris by natural or human agencies. Nevertheless, this approach has evolved rapidly during the last decade with the advent of digital terrain models that make it possible to link geologic processes to time-evolved changes in topography (Hancock et al., 2003).

5.5.4   Empirical Approach

Based on direct measurement of erosion and slope evolution, often in the laboratory and sometimes in the field. Haigh (1978) embraced this approach (using erosion pins) to study the evolution of mine-spoil banks in Wales. He noted a major problem with this approach, namely, the enormous times often required to see measurable changes. Nash (1977) cleverly circumvented this temporal problem by measuring the profiles of a series of ancient coastal bluffs (whose geological ages were known) that had been formed when lake levels were higher. He then compared these profiles (and their changes over time) with the geometry (profiles) of the modern wave-cut bluffs.

5.6   ANTHROPOGENIC SLOPES AND LANDFORMS

When discussing landforms and landform evolution, it is important to distinguish between natural and man-made (anthropogenic) forms. The materials comprising these artificial landforms and their morphology may be quite different than those of natural landforms. Both types of landforms might be subjected to the same geologic processes, but differences in morphology and composition could affect how they respond to these processes and their relative stability over time.

Nevertheless, earthen anthropogenic landforms, for example, earth dams, rock waste piles, and so forth, should respond in a similar manner as natural landforms when subjected to the same geologic processes. In fact, Hancock et al. (2003) argue that natural landforms provide analogues for postmining landscapes and rehabilitation designs in terms of landscape stability and blending rehabilitated structures with the surrounding natural landscape.

Just how important are anthropogenic landforms? Haigh (1978) claims that humans have become an important geomorphic agent and that a large percentage of the earth's landforms are man-made and artificial (anthropogenic). He classifies the processes leading to these artificial landforms in Table 5.1.

5.7   SLOPE EVOLUTION AND LONG-TERM STABILITY

5.7.1   Evolution and Morphometry of Spoil Mounds

Haigh (1978) observed that profiles of older spoil mounds and colliery tips in Wales had compound profiles (concave at bottom and convex at top), whereas freshly dumped and compacted spoil lacked this convexity at the top (see Figure 5.2).

The top of the mound tended to be an area of devegetation and active erosion, producing a convex profile over time. Haigh also measured rates of erosion using soil pins at different positions on the slopes of strip-mine spoil

TABLE 5.1   Classification of Anthropogenic Landforming Processes

Direct Anthropogenic Processes	Indirect Anthropogenic Processes

Constructional
  tipping: loose, compacted, molten
  grading: molded, ploughed, terraced
	Acceleration of erosion and sedimentation
	  agricultural activity and clearance of vegetation
	  engineering, especially road construction and
	    urbanization
	  incidental modifications of hydrologic regime

Excavational
  digging, cutting, mining, blasting of
    noncohesive materials, cratering,
    trampling, churning
	Subsidence: collapse and settling
	  mining
	  fluid withdrawal from the ground
	  thermokarsting

Hydrological interference
  flooding, damming, canal construction
  dredging, channel modification
  draining
  coastal protection
	Slope failure: landslides and flows
	  loading
	  undercutting
	  shaking
	  lubrication

	Earthquake generation
	  loading (reservoirs)
	  lubrication (deep well injection)

Source: From Haigh, 1978.

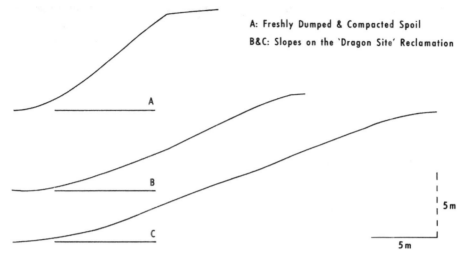

A: Freshly Dumped & Compacted Spoil

B&C: Slopes on the 'Dragon Site' Reclamation

Figure 5.2   Modern spoil-bank slope profiles showing change in shape with time and development of convex top (after Haigh, 1978).

banks near Henryetta, Oklahoma. He observed accretion at the toe and erosion over the rest of the slope, as shown in Figure 5.3. As a result, the spoil-bank slopes in Oklahoma tended to develop a compound profile similar to those observed in the Welsh spoil mounds (see Figure 5.2). The greatest amount of ground retreat (erosion) occurred on sharp-crested divides and least on convex divides.

From these studies and observations, Haigh (1978) reached the following conclusion about the best way to construct such spoil mounds:

> The risk of devegetation [and erosion] might be reduced if slopes were, in the future, designed with well rounded upper convexities which would be less liable to morphological alteration.

5.7.2   Evolution and Morphometry of Natural Slopes

Nash (1977) investigated the morphological evolution of wave-cut bluffs along the coastal margins of Lake Michigan. The modern wave-cut cliffs consist mostly of stratified soil deposits of glacial materials including clay tills and outwash sands. The bluffs start out initially with linear, planar slope faces. Nash was interested in determining their ultimate equilibrium profile and also in identifying a morphological model that best described the evolution and eventual equilibrium shape of this profile.

A number of two-dimensional, slope-evolution, or morphological models have been proposed that predict spatial changes or, more precisely, the rate of change of elevation (dY/dT) of any point on a slope, with elapsed time

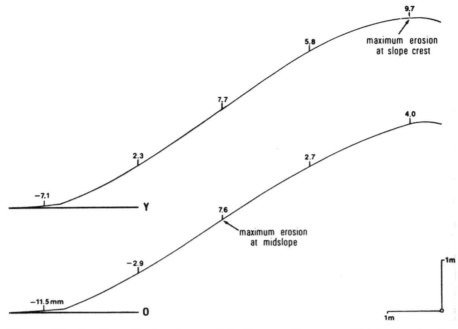

Figure 5.3 Erosion of strip-mine spoil banks near Henryetta, Oklahoma, showing compound slope profile development (convex at top, concave at bottom) with time. Numbers on stakes indicate depth of erosion in millimeters (from Haigh, 1978).

(T) and coordinate locations (X, Y). Examples of these models include the following:

Model No. 1 $dY/dT = -A$
Model No. 2 $dY/dT = -B(dY/DX)$
Model No. 3 $dY/dT = -C$ (height above base)
Model No. 4 $dY/dT = -D$ (distance from from crest)^n
Model No. 5 $dY/dT = -E$ (d^2Y/d^2X)

Graphical simulations of these models are shown in Figures 5.4–5.8. Each of the models has some physical basis. Model No. 2, for example, describes the "parallel retreat of slope" concept, which postulates that, upon reaching its limiting slope angle (angle of repose), a slope retreats back at a constant inclination. A purely frictional, sandy slope whose stability is independent of slope height could conceivably fit this model. Model No. 4 fits observations from the Universal Soil Loss equation, which indicates that rainfall-erosion losses from a slope (all other factors equal) are a function of the slope length. Model No. 5 is the so-called diffusion model, which postulates that, in a

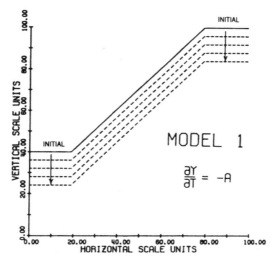

Figure 5.4   Evolution of hillside slope when rate of lowering is uniform over entire slope profile (from Nash 1977).

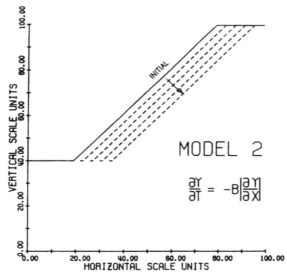

Figure 5.5   Evolution of hill slope when rate of lowering at point on slope is proportional to profile gradient at point (from Nash 1977).

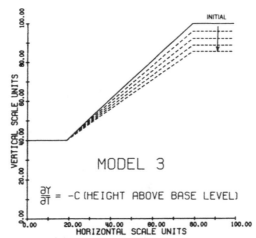

Figure 5.6   Evolution of hillside slope when rate of lowering of a point on slope is proportional to elevation of point (from Nash 1977).

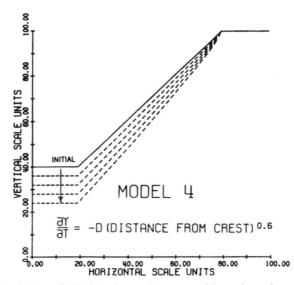

Figure 5.7   Evolution of hillside slope when rate of lowering of a point on slope is proportional to distance that point lies from crest or divide (from Nash 1977).

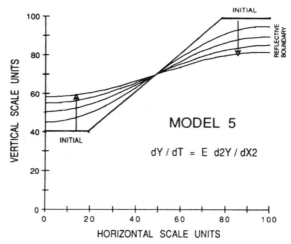

Figure 5.8 Evolution of hillside slopes when rate of lowering of point on slope profile is proportional to profile curvature at that point, assuming reflective left and right boundaries (from Nash, 1977).

transport-limited slope, the passage of material down the slope from a point above is limited by the transfer rate at a point below. The slope profile in this case adjusts itself over time to optimize this stepwise or sequential transfer of material downslope by various erosion or mass-wasting processes. Note that in the diffusion model, an initially planar slope evolves over time into a compound (concave-convex slope) as shown in Figure 5.8.

Nash (1977) tested the applicability of these morphometric models by first ascertaining the time-evolved slope profiles of the initially planar, wave-cut bluffs. To do this, he examined the slope profiles of the present-day, modern wave-cut bluffs along Lake Michigan and those of ancient, abandoned bluffs marking former glacial-lake margins. The study assumed that slope processes at work on the bluffs have remained relatively constant over geologic time. The ancient bluffs and their ages respectively, are the Nipissing bluffs (4,000 years old) and Algonquin bluffs (10,500 years old). The bluffs are easily accessible and clearly visible at a slightly higher elevation than their modern-day counterparts. By resorting to this technique, Nash was able to bypass one of the main limitations of the morphometric approach, namely, the huge amount of time normally required to observe measurable and meaningful changes in slope profiles.

Actual slope profiles for these three bluffs superposed at their midpoint are shown in Figure 5.9. The correspondence (or fit) between the profiles, predicted by the diffusion model and the actual profiles, was examined for various diffusion constants. The configurations predicted by the diffusion model for an abandoned bluff after 4,000 years and 10,500 years, using a diffusion

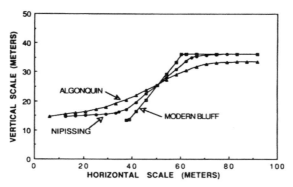

Figure 5.9   Modern bluff profile, Nipissing bluff profile (4,000 years old), and Algonquin bluff profile (10,500 years old) superposed at their midpoints (from Nash, 1977).

coefficient of 0.012 m²/yr and an initial, planar profile similar to the profile of the modern bluff are shown in Figure 5.10. According to the diffusion model, the slope profiles gradually change over time from a linear to a compound (concave-convex) configuration, as illustrated in Figure 5.10.

The fit (or correspondence) between actual and predicted profiles is quite good, as can be seen by comparing slope profiles in Figure 5.9 with those in Figure 5.10. This modeling work also shows that in transport-limited slopes, at least, a planar slope with constant inclination, typical of most conventional grading practice, is not a not a stable, long-term equilibrium slope.

5.7.3   Effect of Climate on Hillslope Form

Considerable controversy has swirled about the influence of climate on hillsope form. King (1953, 1957, and 1962) maintained that climate does not

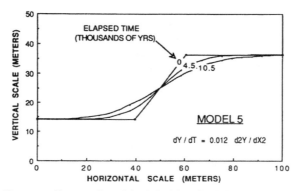

Figure 5.10   Slope profiles predicted by Model 5 for initial planar slope after 4,000 and 10,500 years of elapsed time using diffusion coefficient of 0.012 m2/yr and initial inclination similar to present wave-cut bluff (from Nash, 1977).

play a direct controlling force and that similar landforms may be found under like conditions of bedrock and relief in all climatic environments short of glaciation or wind-controlled, sandy deserts. This viewpoint was espoused explicity as a "canon" in his 1953 paper (p. 748) as follows:

> "There is a general homology between all epigene landscapes. The differences between landforms of humid-temperate, semiarid, and arid environments are differences only of degree."

A somewhat different view was presented by Toy (1977), who claimed that climate affects hillslope evolution via its influence on the intensity and type of geologic processes, such as rainfall erosion and mass wasting. Toy (1977) investigated the relationship between hillslope form and climate by selecting 29 hillslope sites along two traverses: one from Berea, Kentucky, to Pioche, Nevada, and the other from Glendive, Montana, to Ruidoso, New Mexico. These two traverses spanned a range of climatic regimes ranging from humid to arid. A rigorous site-selection procedure was adopted to minimize the importance of nonclimatic factors, for example, composition, structure, and so forth.

Results of Toy's research appears to support the hypothesis that hillslopes reflect climate variations. Hillslopes in arid regions tend to be shorter, steeper, and have a smaller radius of curvature of the convex segment than those in humid areas. This generalization is illustrated schematically in Figure 5.11. A consequence of this hillslope geometry is that arid regions should have higher drainage densities and stream frequencies than humid ones.

Regression analyses of study results indicated that spring and summer precipitation, potential evapotranspiration, and water availability are the aspects of climate most closely associated with hillslope form. Summer and spring appeared to be the seasons when precipitation had the greatest influence on hillslope form.

Dunkerley (1978) has drawn attention to some serious flaws in Toy's experimental design and statistical interpretations. According to Dunkerley, Toy

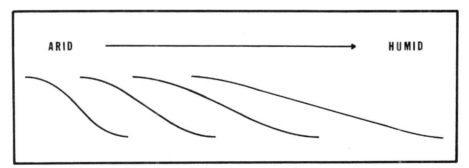

Figure 5.11   Conceptualized hillslope profiles in arid and humid climates (from Toy, 1977).

did not identify all of the variables that may act separately to control hillslope profile form—e.g., hillslope plan form and evolutionary stage. With regard to the latter, Dunkerley noted a discrepancy between the length of the climatic record (1951–1971) used in Toy's analysis and the likely period of time over which the hillslopes had actually developed. Nevertheless, Toy's research provided some evidence for the existence of quantifiable relationships between geomorphological processes and hillslope forms. His findings presage later research in which digital terrain models were coupled with geologic-process equations to predict landscape evolution.

5.8   DIGITAL TERRAIN MODELS

5.8.1   Salient Characteristics of Digital Terrain Models

Digital terrain models allow any landscape (or landform) to be converted to a spatial coordinate (x, y, z) system and to be transformed into a grid of digital elevations (Moore et. al., 1988). Examples of digital elevation models and their corresponding topographic contour plans are shown in Figure 5.12. The file containing the spatial (x, y, z coordinates) can be linked with geologic-process equations so that time-evolved landforms can be evaluated. This permits estimates of surface lowering, soil loss, and gully incision at different elapsed times or different ages (Hancock and Wilgoose, 2002; Evans et al., 2000; Evans and Willgoose, 2000; Willgoose et al., 1991).

Hancock et al. (2003) note that digital models make it possible to evaluate landscape stability over the short (annual), medium (hundreds of years), and long term (thousands of years). These different time periods are relevant to different types of problems, for example: "short" for most engineering projects and sites with critical structures; "medium" for hazardous-waste landfills and toxic mine tailings or spoil heaps; and "long" for high-level, nuclear-waste disposal fills. Most conventional slope-stability analyses do not provide these time dependent predictive features.

Digital terrain models offer the following important opportunities and advantages:

- First and higher order landforms can be represented and evaluated by adjusting the range of scales from grid spacings of less than a meter to grid scales of tens of meters (Hancock et al., 2002; Ibbit et al., 1999).
- Topographic attributes, such as local ground slope and drainage area per unit contour length, can be determined from the digital terrain model and used to define quantitatively the spatial variation in dominance of different erosional processes (Dietrich et. al., 1992, 1993) and to identify land parcels (areas) most susceptible to shallow landsliding and/or channel initiation (Montgomery and Dietrich, 1989, 1994).

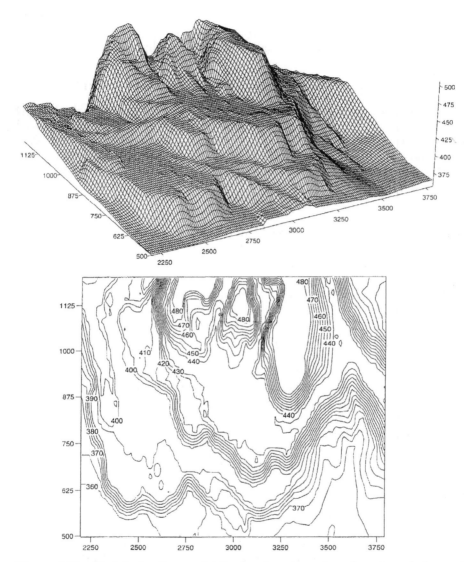

Figure 5.12 Digital elevation model (top) and contour plan of waste rock dump at Parraburdo, Australia (after Hancock et. al., 2003).

- Time-evolved landform analysis makes it possible to evaluate different embankment or slope designs from simple linear slopes, benched and/or bermed slopes, to compound slope shapes (Hancock et al., 2003).

5.8.2   Example of a Linked, Digital Terrain Model—SIBERIA

SIBERIA is a physically based predictive model (Willgoose et al., 1991A) that can simulate the geomorphic evolution of landforms subjected to fluvial erosion and mass-transport processes. SIBERIA links widely accepted hydrologic and erosion process models to changes in topography over time.

The sediment transport equation of SIBERIA consists of two terms:

$$q_s = q_{sf} + q_{sd} \tag{5.1}$$

where q_s = the sediment transport rate per unit width
q_{sf} = the fluvial or rill sediment transport
q_{sd} = the diffusive or inter-rill transport

The fluvial sediment transport term (q_{sf}) is based on the Einstein-Brown model, which is:

$$q_{sf} = \beta_1 q^{m_1} S^{n_1} \tag{5.2}$$

where q = the discharge per unit width
S = the slope (gradient) in steepest downslope direction
β_1, m_1, n_1 = model parameters

The diffusive transport term (q_{sd}) is expressed as:

$$q_{sd} = DS \tag{5.3}$$

where D = diffusivity
S = the slope (gradient)

In the SIBERIA digital terrain model, the diffusive term models inter-rill erosion (i.e., raindrop splash and sheet flow), creep, and landsliding. Use of the model requires calibration to determine the model parameters (β_1, m_1, n_1); an appropriate value for the diffusivity (D) must also be determined. The calibration procedure, which is quite complex, is described in detail by Willgoose and Riley (1998).

Essentially the model calibration process adjusts the model parameters so that the behavior of the model is consistent with observed behavior in the field.

Calibration can be accomplished by using either laboratory or field data. The best and most reliable method of calibrating SIBERIA is to use long-term hydrology and sediment data measured from field catchments (Evans and Willgoose, 2000A; Hancock et al. 2000). Unfortunately, acquisition of field data is often expensive and time consuming. Instrumentation must be installed to monitor catchment sediment loads and hydrologic flows. An alternative procedure is to use rainfall simulation to accelerate the calibration process. This requires several actions: (1) transporting a rainfall simulator to the site of interest, (2) constructing an artificial catchment, (3) subjecting the surface to a range of simulated rainfall events, and (4) collecting runoff and sediment from the catchment. Rainfall simulations can deliver rainwater over a range of intensities and slopes, thereby speeding up the calibration process considerably.

For long-term elevation changes, the usual procedure is to model the average effect of the various slope processes over time. Therefore, individual events are not modeled; instead, the average effect of many aggregated events over time are modeled. This procedure allows the landform that is being modeled to evolve in response to erosion, and the SIBERIA output describes how a catchment or landform is expected to look, on average, at any given time.

SIBERIA can be adapted to particular requirements and conditions whether it be a visual inspection of an evolving landform, on the one hand, or estimate of soil-erosion losses and sediment delivery, on the other. In many cases, it may be unecessary to calibrate the erosion model specifically for the erodible material underlying a landform. In this case, different erodibilities can be arbitrarily assigned to surface materials, and SIBERIA can then be used to examine the effect of these differences, along with climate, on landscape stability. In this way, best- and worst-case scenarios of potential designs can be predicted.

Generic parameters can also be introduced to provide a sensitivity analysis of different designs of artificial landforms or erosion-control structures (Kirby, 1971). For example, the parameters m_1 and n_1 in Equation 5.2, and the relationship between these parameters (see Table 5.2), are known to control whether a landscape will develop either by gullying or slope-wash processes (Kirkby, 1971). Accordingly, if the erodibility of the surface material is known, then an assessment of the erosive behavior or response of the landform can be made by adjusting the values of m_1 and n_1.

TABLE 5.2   Effect of Different Erosion Model Parameters on Erosion Processes

Erosion Parameter		Erosion Process
M_1	N_1	
1	1–2	soil wash without gullying
2	2	soil wash with gullying
>2	>2	gullying

Source: From Kirkby (1971).

Hancock et al. (2003) used the SIBERIA model to simulate the geomorphic development of selected natural catchments in the Tin Camp Creek area in northern Australia over geological time. They varied the value of the diffusivity (D) to test the sensitivity of the model to this parameter. A three-dimensional digital elevation model of one of the study catchments and the SIBERIA simulation are compared in Figure 5.13. Simulations were run with no diffusion ($D = 0$) and with diffusion approximately three times the calibrated value ($D = 0.1$), as shown in Figure 5.14.

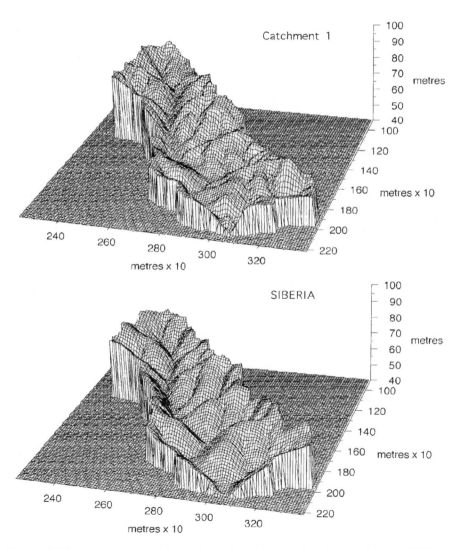

Figure 5.13 Comparison of three-dimensional digital elevation model and SIBERIA simulation of study catchment (after Hancock et al., 2003).

D = 0

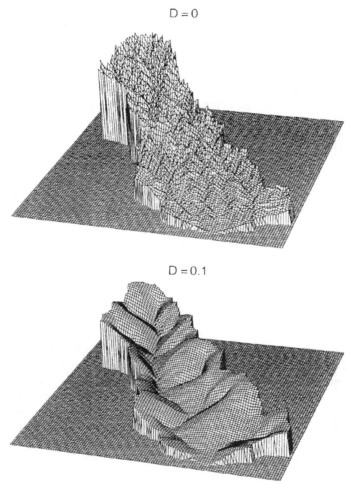

D = 0.1

Figure 5.14 SIBERIA model simulation of study catchment with differing diffusity values (after Hancock et. al., 2003).

The absence of diffusion ($D = 0$) produced a landscape that had an angular or jagged surface that was dominated by fluvial processes and with short hillslopes. Increased diffusivity (much greater than the calibrated value) produced a landscape surface that was very smooth compared with the observed landscape; it also had much longer hillslopes. These results demonstrate the importance of correctly determining the diffusivity.

5.8.3 Applications of Digital Terrain Modelling

The SIBERIA erosion model has been used to analyze the evolution of natural landforms including catchments and watersheds (Willgoose et. al., 1991A,

1991B; Ibbit et al., 1999; Hancock and Wilgoose, 2002; Hancock et al., 2002). This same technique, can also be employed to evaluate the stability and performance of artificial landforms and postmining landscape designs (Hancock et al., 2003; Evans et.al. 2000; Evans and Willgoose, 2000; Willgoose et al., 1991; Willgoose and Riley, 1998).

Landscape evolution models like SIBERIA enable one to rapidly assess and visualize how different landform designs evolve over time. This means that current landforms or proposed designs for earthen structures, such as waste-rock mounds or tailings stockpiles, can be assessed for their long-term stability. This capability also enables a designer to evaluate the long-term effectiveness of such erosion-control measures as berms, swales, contour benches, and so forth that might be incorporated into a design.

Willgoose and Riley (1998) used SIBERIA to assess the long-term stability of engineered landforms associated with the rehabilitation of the Ranger Uranium mine, Northern Territory, Australia. Regulations require that tailings be contained safely and without deep incision by erosion gullies for periods in excess of 1000 years. Simulations of a preferred above-grade waste-rock dump after elapsed times of 0, 500, and 1,000 years are shown in Figure 5.15. Preliminary analysis of the preferred option indicated that erosion on the order of 7 to 8 meters would occur on the structure in a period of 1,000 years. This depth of erosion (or incision) may be sufficient to compromise the integrity of containment.

With regard to the influence of various erosion-control measures on artificial landforms, Hancock et al. (2003) have studied the effect of berms and contour benches on long-term stability. Figure 5.16 demonstrates the ability of SIBERIA to model erosion and gully development on the face of a field-scale tailings dam. Gullies were triggered by the concentrated flow of water on the flat top surface of the dam and were initiated by the transition from the flat top to the steep slope below. Runoff is captured on the flat cap and directed down the face of the slope where it incises into the surface. The same problem would exist in the case of a slope with contour benches. The SIBERIA simulation suggests that the benches and contour breaks might be effective at managing short term erosion—provided the benches are maintained. But in the long run the benches will be ineffective and actually promote gully incision once maintenance ceases.

5.8.4 Design of Stable Landforms

Experience with landscape-evolution models like SIBERIA provides insights about ways that landform design can minimize erosion losses and maximize long-term stability. Other landscape-evolution models that have been developed have similar capabilities. A software program developed by Bugosh (2005) applies fluvial geomorphic principles to upland landform designs. This program helps the user create a landscape design that mimics the functions of a natural landscape, one that would evolve naturally over time and be in

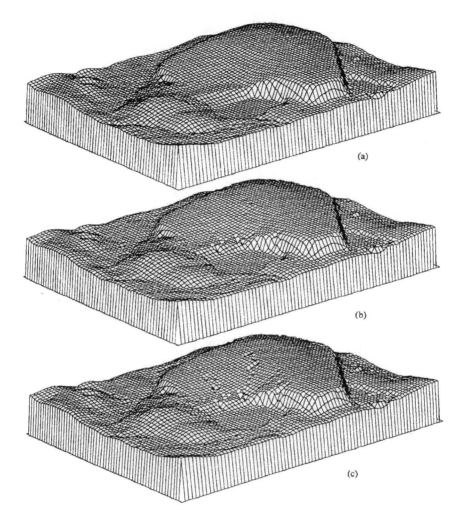

Figure 5.15 SIBERIA simulations of proposed above-grade waste-rock dump at Ranger Uranium Mine, Northern Territory, Australia showing elevations at elapsed times of 0, 500, and 1,000 years (after Willgoose and Riley, 1998).

stable hydrologic equilibrium. Although anthropogenic landforms (e.g., dams, embankments, cut-and-fill slopes along highways, pads for hillside housing, rock-waste dumps, etc.) are frequently constructed with uniform, planar slope faces, few slopes in nature are planar and uniform. Furthermore, evidence suggests (Gyasi-Agyei, 1996) that in soil-mantled landscapes, linear slopes will erode to a concave profile over time, thus increasing the sediment loss of the profile until a concave profile has been achieved.

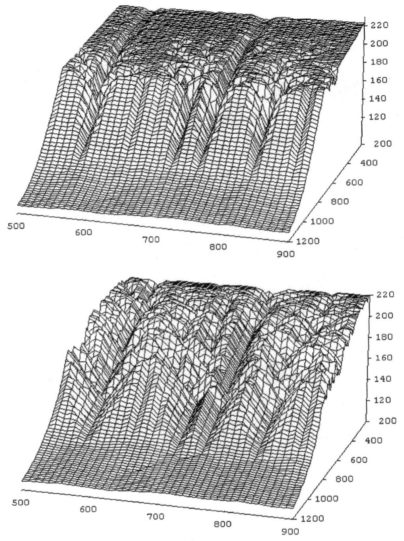

Figure 5.16   SIBERIA simulations of gully development on tailings dam face at 1 minute (top) and 6 minutes (bottom) (after Hancock et al. 2003).

Similar observations based on much earlier field and laboratory studies (Meyer and Kramer, 1969; D'Souza and Morgan, 1976; Young and Mutchler, 1969) also show that slopes with concave profiles generate less sediment than those with a uniform, linear profile. These findings are reviewed in detail in Chapter 4. These findings have prompted suggestions that linear slopes be eschewed in favor of curvilinear slopes similar to those found in nature

(Haigh, 1978; Schaefer et al., 1979; Schor, 1992) to reduce soil losses and maximize long-term stability in addition to enhancing visual appearance.

Contour benches are routinely used in conventional designs of artificial landforms to reduce erosive potential. However, very few linear slopes with benches are observed in nature; furthermore, simulations with landscape-evolution models like SIBERIA indicate that over the long term, benches and contour breaks are unreliable (Gyasi-Agyei, 1996; Hancock et al., 2003), because they tend to channel water in concentrated flow paths, leading to severe gullying with time.

5.9 REFERENCES

Bugosh, N. 2006. Primer to the GeoFluv™ Method for Reclamation Landform Design. Paper (unpublished) presented at the National Interactive Forum on Geomorphic Reclamation. Sponsored by the Office of Surface Mining, Western Region, Farmington, N.M., Sept 12-14, 2006.

Davis, W. M. 1899. The geographical cycle. *Geography Journal* 14:481–504.

Dietrich, W. E., C. J. Wilson, D. R. Montgomery, J. MacKean, and R. Bauer. 1992. Erosion thresholds and land surface morphology. *Geology* 20:675–679.

Dietrich, W. E., C. J. Wilson, D. R. Montgomery, and J. MacKean. 1993. Analysis of erosion thresholds, channel networks and landscape evolution using a digital terrain model. *Journal of Geology* 101:259–278.

D'Souza, V. P., and R. P. Morgan. 1976. A laboratory study on the effect of slope steepness and curvature on soil erosion. *Journal of Engineering Research* 21:21–31.

Dunkerly, D. L. 1978. Climatic geomorphology and fully developed slopes: A discussion. *Catena* 5:79–80.

Evans, K. G., M. J. Saynor, G. R. Willgoose, and S. J. Riley. 2000. Post-mining landform evolution modelling: 1. Derivation of sediment transport model and rainfall-runoff model parameters. *Earth Surface Processes and Landforms* 25(7): 743–763.

Evans K. G., and G. R. Willgoose. 2000. Post-mining landform evolution modelling: 2. Effects of vegetation and surface ripping. *Earth Surface Processes and Landforms* 25(8):803–823.

Fookes, P. G., and P. R. Vaughan (eds.). 1986. *A Handbook of Engineering Geomorphology.* New York: Chapman and Hall.

Gyasi-Agyei, Y., and G. R. Willgoose. 1996. Evaluation of the use of contour banks as a post-mining rehabilitation control option using a digital terrain based rainfall runoff erosion model. In *Application of Geographic Information Systems in Hydrology and Water Resources Management,* eds. H. Holzman and H. P. Nachtnebel, 143–150. Vienna Austria: HydroGIS '96, 16–19 April.

Haigh, M. J. 1978. Evolution of slopes on artificial landforms, Blaenavon, United Kingdom. Research Paper No. 182., University of Chicago, Department of Geography.

Hancock, G. R., and G. R. Willgoose. 2002. The use of a landscape simulator in the validation of the SIBERIA landscape evolution model: Transient landforms. *Earth Surface Processes and Landforms* 27(12):1321–1334.

Hancock, G. R., G. R. Willgoose, and K. G. Evans. 2002. Testing of the SIBERIA landscape evolution model using the Tin Camp Creek, Northern Territory, Australia, field catchment. *Earth Surface Processes and Landforms* 27(1):25–143.

Hancock, G. R., R. Loch, and G. R. Willgoose. 2003. The design of post-mining landscapes using geomorphic guidelines. *Earth Surface Processes and Landforms* 28:1097–1110.

Ibbit, R. P., G. R. Willgoose, and M. J. Duncan. 1999. Channel network simulation models compared with data from the Ashley River, New Zealand. *Water Resources Research* 35(12):3875–3890.

King, L. C. 1953. Canons of landscape evolution. *Geological Society of America Bulletin* 64(7):721–752.

———. 1957. The uniformitarian nature of hillslopes. Edinburgh Geological Society Transactions 17:81–102.

———. 1962. *Morphology of the Earth.* Edinburgh, Scotland: Oliver and Boyd.

Kirkby, M. S. 1971. Hillslope process-response models based on the continuity equation. In *Slopes, Forms and Process.* London: Institute of British Geographers Serial Publication No. 3.

Meyer, L. D., and L. A. Kramer. 1969. Erosion equations predict land slope development. *Agricultural Engineering* 50:522–23.

Montgomery, D. R., and W. E. Dietrich. 1989. Source areas, drainage density, and channel initiation. *Water Resources Research* 25(8):1907–1918.

Montgomery, D. R., and W. E. Dietrich. 1994. A physically based model for the topographic control on shallow landsliding. *Water Resources Research* 30(4):1153–1171.

Moore, I. D., E. M. O'Loughlin, and G. J. Burch. 1988. A contour-based topographic model for hydrological and ecological studies. *Earth Surface Processes and Landforms* 13:305–320.

Nash, D. B. 1977. The evolution of abandoned, wave-cut bluffs in Emmet County, Michigan. PhD dissertation, University of Michigan, Ann Arbor.

Penck, W. 1953. *Morphological Analysis of Land Forms.* London: MacMillan.

Schaefer, M., D. Elifrits, and D. J. Barr. 1979. Sculpturing reclaimed land to decrease erosion. Paper presented at the Symposium on Surface Mining Hydrology, Sedimentology and Reclamation, University of Kentucky, Lexington, Kentucky, December 4–7.

Sharp, R. P. 1982. Landscape evolution (a review). *Proceedings of the National Academy of Science* 79:4477–4486.

Schor, H. J. 1992. Hills like nature makes them. *Urban Land* (March):40–43.

Schor, H. J., and D. H. Gray. 1995. Landform grading and slope evolution. *Journal of Geotechnical Engineering (ASCE)* 121(GT10):729–734.

Toy, T. T. 1977. Hillslope form and climate. *Geological Society of America Bulletin* 88: 16–22.

Willgoose, G. R., R. I. Bras, and I. Rodrigueez-Iturbe. 1991A. A physically based coupled network growth and hillslope evolution model: 1. Theory. *Water Resources Research* 27(7):1671–1684.

————. 1991B. A physically based coupled network growth and hillslope evolution model: 2. Applications. *Water Resources Research* 27(7):1685–1696.

Willgoose, G. R. 1994. A physical explanation for an observed area-slope-elevation relationship for catchments with declining relief. *Water Resources Research* 30(2): 151–159.

Willgoose, G. R., and S. Riley. 1998. The long-term stability of engineered landforms of the Ranger Uranium Mine, Northern Territory, Australia: Application of catchment evolution model. *Earth Surface Processes and Landforms* 23(3):237–259.

Young, R. A., and C. K. Mutchler. 1969. Effects of slope shape on erosion and runoff. *Transactions of the American Society of Agricultural Engineers* 12:231.

6 Hillside Grading Fundamentals

6.1 INTRODUCTION

The purpose of this chapter is to present an overview of the fundamentals and to describe important details of the grading process in a compact and concise format. The intent here is to provide the reader with the necessary background and to set the stage for some of the issues discussed in later chapters. It is not meant to be comprehensive or the final word on the subject, as there can be many variations and deviations in method based on given site conditions and unique circumstances. Instead, the information in this chapter is designed to be general yet informative in nature; it is also based on the principal author's more than 30 years of experience in the field.

Grading is a major undertaking, particularly with the scale of today's sites—e.g., planned hillside community development, surface mining, landfills, and other applications. Many elements and operations combine to make up the grading process. Both the outline and description that follows are tailored to the more complex operations in hillside areas for land development purposes. Constraints are greater, and space is more limited in hillside areas, because every square foot of graded land has a very high value.

6.2 PURPOSE OF GRADING

Whenever a change in land use necessitates alteration of the underlying topography, grading is the technique by which this change is accomplished. Some form of slope grading is typically required for construction and reclamation purposes. Traditional grading entails preparing an uneven or sloping surface so that it can accommodate building sites and roadways. In addition to creating level or flat surfaces, grading practice must allow for drainage and conveyance of surface runoff and take into account possible geologic hazards such as landslides, adverse bedding conditions, faulting, surficial erosion and potential subgrade problems, for example, weak or compressible soils, shrink-swell instability, and so forth. An effective grading design should prevent or minimize potential off-site damages from runoff and sediment discharge to adjacent downslope properties. Finally, a good grading design should be visually attractive and blend into the natural topography as much as possible.

Grading practice associated with land disposal of overburden spoil or land reclamation has slightly different requirements. The construction of level building pads is not a major goal of the grading design in this case. Here, the emphasis is more on creating a stable surface, or landform, that is visually compatible with the surrounding landscape, that can be re-vegetated effectively, and that provides for the safe conveyance of runoff at noneroding velocities.

6.3   GRADING CONSIDERATIONS

6.3.1   Major Stakeholders

Grading is a complex, multifaceted activity that requires careful planning and coordination from a number of different disciplines, in particular geology and soils engineering but also land planning, various facets of civil engineering, phase I and II environmental studies, and earthwork contracting.

A number of regulatory agencies with responsibility for planning and environmental enforcement also have a stake in grading practice and decisions. It almost always requires addressing archaeology, paleontology, fluvial morphology, biological habitat, and endangered species issues, "blue-line stream" considerations, Native American, and other historical issues. These considerations may bring state, federal, and local jurisdictions and their requirements to bear before any grading can be conducted. Grading is also time sensitive as state, federal, and local agencies will have restrictions on when such grading can take place and under what conditions, because it can impact nesting seasons and migratory bird issues.

Various agencies that have a stake in the above specialty fields include the following:

- *State entities:* State Departments of Fish and Game charged with the protection of endangered species and their habitat or the application of mitigation measures in the case of approved takings. State Uniform Building Codes and Standards that regulate basic grading standards and safety measures.
- *Federal entities:* Army Corps of Engineers and U.S. Fish and Wildlife Service, both of which control the limits of grading, the extent of "taking," and the establishment and monitoring of short- and long-term mitigation requirements. In addition the Army Corps is responsible for blue line stream protection and sets mitigation requirements where taking is approved.
- *Local entities:* Grading Ordinances, Hillside Development, and Management Ordinances.

6.3.2   Selection of Grading Equipment

Large scale earthmoving and grading projects require the use of specialized equipment with particular capabilities. This requirement holds true regardless of whether conventional or landforming techniques are employed. Different types of key equipment used in grading and earthwork operations are briefly described in this section.

A. *The Bulldozer*

This grading machine is of critical importance in preparing the ground for future cutting and filling operations. A bulldozer typically has two components—a blade mounted on the front to cut and push dirt and ripper teeth, either one or two, mounted on shanks at the rear (see Figure 6.1). The latter are designed to rip and loosen soil and bedrock. Bulldozers come in a variety of sizes and horsepower, each designed for specific tasks or assignments. Bulldozers are capable of performing fairly intricate and precise construction operations. To optimize their productivity for large-scale work that requires this level of precision, the latest machines have been equipped with GPS guidance systems.

Bulldozers perform a number of tasks that are characteristic of grading operations. They perform the initial pioneering, access-road building, clearing, and grubbing. Bulldozers remove surface topsoil and

Figure 6.1   Grading equipment: track dozer equipped with blade, ripper teeth, and slope board at work.

colluvium and, where required, create benches in the underlying bed-rock.

To cut a slope, a so-called "slope board" is installed on one side of the dozer, the angle of which can be adjusted to meet the required inclination of the cut. Dozers further prepare the cut areas, using blade and ripper teeth to facilitate scooping up the dirt with the scrapers. In bedrock conditions, a dozer's twin-ripper teeth can often break up the bedrock, depending on the rock's degree of hardness and how well the rock is cemented. This capability also depends upon the size of the dozer. When bedrock is much harder, a single shank ripper tooth is used. A single tooth can handle and manipulate harder material more effectively and exploit fractures in the bedrock by breaking it up with-out the need of blasting.

B. *The Scraper*

The scraper is the primary earthmoving machine; it hauls dirt from a cut to a fill site and places the material in this new location. Because of their weight, these machines act simultaneously as compactors of the fill during the placement stage. Scrapers are built as single-engine machines, with a motor up front only, pulling it. With twin-engine scrapers, a second motor in the rear provides push power at the same time (see Figure 6.2). Twin engine-scrapers are typically more cost-efficient, particularly in hilly terrain that involves fully loaded uphill hauls.

Figure 6.2 Grading equipment: twin-engine 657 scraper (32 cubic yard capacity) fully loaded on uphill haul.

C. *Water Truck*

Water trucks (see Figure 6.3) provide essential water that is needed in most grading operations to provide for proper compaction. The water content during compaction controls the density and other engineering properties of compacted soil, such as strength and hydraulic permeability. The influence of water content on compacted soil properties is discussed in detail in Chapter 3 (Section 3.6.5, Optimizing Compaction). Water, in specified amounts, is the agent that allows soil particles to move into a denser packing configuration when the soil is compacted under the action of rollers and other types of compaction machinery.

D. *Compaction Equipment*

Compaction machinery typically consist of rollers with either smooth or rough, irregular surfaces. The so-called sheepsfoot roller typically consists of two side-by-side steel drums that have 6-inch steel teeth, or "feet," projecting from the drum (see Figure 6.4). This device exerts a kneading compaction effort that aids in densification of soils, particularly clayey soils. Sheepsfoot rollers are frequently used to compact fills that are placed horizontally in layers or "lifts" as well as on slopes to compact their outer surface. In the latter case, the drums are lowered by a bulldozer on a cable and are moved up and down the slope face as shown in Figure 6.4.

The sheep-foot-like, 6-inch projections on the rollers initially penetrate into the soil when the drums roll over the ground. As the soil compacts during successive passages, the feet penetrate less and less

Figure 6.3 Grading equipment: water truck.

Figure 6.4   Grading equipment: sheepsfoot roller.

and the roller gradually lifts up and travels on the tips of the feet. This action, which is referred to as "walking out," provides a good visual indication of the degree or extent of compaction. When absolute smoothness and extra compaction is desired on a slope face, a grid roller drum (see Figure 6.5) is used instead of (or in addition to) a sheepsfoot roller.

All of the above equipment can often be seen working together on a project site, as shown in Figure 6.6.

6.3.3   Importance of Subsurface Conditions

Most of the conditions that will be encountered during the grading process exist below the surface. Unfortunately, our knowledge of these subsurface conditions is generally limited and normally based on interpretation of a series of exploratory boreholes and shallow-trench excavations. While geotechnical science has advanced greatly in the past two decades, it is often as much an art as a science. Accurate interpretation of subsurface data gleaned from finite holes and trenches, in combination with careful analysis of stereoscopic photos and observation of geomorphic or topographic features, still relies to a great extent on the experience and judgment of a qualified geologist or geological engineer. However, even with the best predictive techniques, Mother Nature never fails to surprise us. It is not unusual, therefore, that unforeseen conditions can emerge when the ground is "opened up" and that more often than not the project will require design corrections or additional mitigation measures. Examples of these "surprises" include:

Figure 6.5 Grading equipment: slope grid roller.

Figure 6.6 Grading project underway showing variety of grading equipment in action.

A. *Unanticipated seepage and springs.* The source may be highly localized in buried aquifer deposits or pockets of ancient perched water.

B. *Asymmetrical alluvial deposits in channels.* Thickness and width of the deposit may vary along the centerline of the channel. Paper calculations often do not reflect this variation.

C. *Wet or saturated soils at depth.* These buried deposits may require special removal and drying before final placement.

D. *Adverse bedding conditions.* Weaker bedrock at depth or strata that dip into excavations can cause difficulties during grading.

E. *Buried boulders.* Undetected boulders, particularly in glacial soils, can give false indications of bedrock in drill holes and pose serious problems regarding disposal and cost issues during excavations.

6.4   ELEMENTS OF HILLSIDE GRADING

Grading consists of both preparatory operations and actual earthmoving activities. A site must be reconnoitered and prepared in advance of actual earthmoving to ensure that earthmoving can proceed effectively and to prevent future problems. Conventional earthmoving, basically, consists of cutting, excavating, or filling operations. The resulting landforms, namely, cuts, slopes, and fills have certain geotechnical attributes that make their behavior quite different from one another.

6.4.1   Preparatory Operations

6.4.1.1   Clearing and Grubbing: Plant cover must be stripped and root structures removed to remove organic or compressible material. Incorporation of organic slash or debris into earthen fills or embankments can lead to serious problems with settlement and zones of weakness, as the organic material decomposes.

Clearing or grubbing can be carried out in an incremental or sequential fashion to avoid leaving large areas of hillslope in a denuded and vulnerable condition to surficial erosion.

6.4.1.2   Preapplication of Water: Areas designated for fills should be considered for prewatering prior to actual earthmoving to avoid subsidence later. Certain type of soils—e.g., loose, high void-ratio silts and clayey silts or soil deposits (alluvium and loess)—tend to collapse when wetted. Exceptionally dry soils, such as those present in upland alluvial fan deposits in arid climates, have low-moisture contents and significant air voids that would normally be occupied by water. These deposits are particularly prone to collapse when wetted or saturated, a phenomenon known as "hydrocompactions." Watering in advance of fill placement (i.e., prewatering) causes the settlement to occur before, as opposed, to after placement of fills and structures.

6.4.1.3   Removal of Deleterious Materials

- *Alluvium and colluvium.* These soils often have high void ratios resulting from the manner in which they are deposited. This loose, high void-ratio condition make these soils relatively weak and compressible. Fills placed upon such soils tend to sink or subside, as the weight of overlying material compresses the soil. Complete removal typically is done to avoid later differential settlement and structural damage. Some older deposits of alluvium may be sufficiently consolidated and/or cemented that they can be left in place. Another alternative in the case of extremely dry, high void ratio alluvium is prewetting (see previous section) and or surcharging to minimize future settlement.
  - High fills in some such cases can provide the surcharge necessary to stabilize and settle the ground if properly monitored and adequate time is allowed for it.
  - Examples of alluvium and colluvium removal are shown in Figures 6.7 and 6.8 respectively.
- *Organic soils.* Peats, mucks, or organic soils (top soils or A-horizon material) should be removed, because they are weak and compressible. A high organic content make them potentially valuable, however, only as a substrate for vegetation establishment. Accordingly, these materials should be saved and stockpiled for later use.

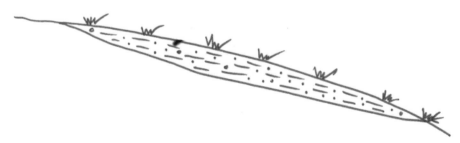

Figure 6.7   Example of colluvium removal.

REMOVAL OF ALLUVIUM (Qal)

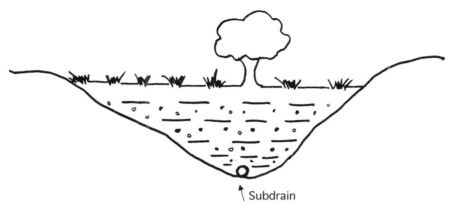

↑ Subdrain

Figure 6.8   Example of alluvium removal.

- *High sulfate content soils.* Sulfates in soils are potentially deleterious to concrete. They should not be allowed to come into contact with concrete foundations; alternatively, a sulfate resistant cement must be specified.

- *Diatomaceous earth.* Diatomaceous earth is a soft, chalklike sedimentary rock that poses special problems with handling, drying, placement, and compaction.

6.4.2   Special Conditions and Precautions

6.4.2.1   Groundwater Removal: High or perched groundwater tables pose a variety of problems during grading operations. Excavation bottoms can become saturated and unstable if groundwater flows upward into an excavation with a sufficient hydraulic gradient. Under these conditions effective or intergranular stresses between soil particles go to zero and a soil loses its shear strength and bearing capacity. Slopes likewise can become unstable if groundwater daylights (exits) at cut faces with a sufficient seepage pressure or hydraulic gradient. A whole array of subsurface drainage systems and measures can be installed (Cedergren, 1997; Forester, 2001; Gray and Sotir, 1996) to intercept and divert groundwater and lower pore water pressures. These measures include horizontal and vertical drain systems, buried subdrains, trench drains, and wicker systems. Examples of a perforated, solid, "burrito" drain system are shown in Figure 6.9.

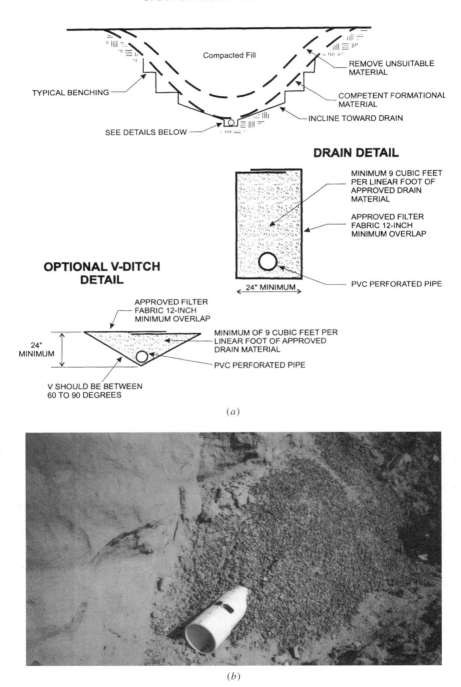

Figure 6.9   Examples of perforated, "burrito"-type subdrain system: (*a*) schematic drawing showing placement and (*b*) photo illustration.

Soils that are excessively wet can also pose a number of problems. For example, wet soils may be too wet to load using scrapers, or they may make the handling and compaction of soils next to impossible. However, there are several techniques available to manage this problem. The measures include the following:

1. *Double Move Operations*

    Where the material is too liquid for scrapers to be able to load, backhoes can top load onto trucks and haul the material to an area where it can be spread, tilled, and adequately dried. After drying, the soil is moved a second time to a location where it is placed and compacted in a fill. This requires not only the very expensive loading and trucking operation (as opposed to the far less expensive scraper haul) but also a double movement.

2. *In Situ Drying of Soils*

    Soils can be dried in place and, then, either used for fill or receive fill, which is placed on top of the wet soil. Landslides are typically caused by adverse groundwater conditions; as a result slide masses frequently contain excess water. Wells and/or hydraulic augurs are often installed in predetermined locations prior to grading to intercept water-bearing strata and pockets. This dewatering procedure improves stability of the slide body and minimizes the chance of uncontrolled failures during grading operations.

3. *Mitigation in Place to Accommodate New Fills*

    An alternative solution has been effectively utilized when the depth of saturated soils is too great, the source of water intrusion is offsite and cannot be kept out, or control of groundwater through a drain system cannot be achieved because of gravity flow limitations. This alternative consists of removing only the upper several feet while a gravel drain blanket, together with a filter layer, is placed over the remaining saturated material that is left in place. Fill is then placed over the top, often well above final grade, to surcharge the site and allow for an adequate settlement period. Settlement markers are installed throughout to measure progression of the consolidation. This could typically take from six to twelve months or longer. Successful implementation depends on whether there is sufficient time for settlement to reach equilibrium before structure construction can begin on the site; but this alternative is by far the least expensive construction technique.

6.4.2.2   Surface Drainage Control: Subsurface drainage is important to insure the mass stability of cuts and fills. Surface drainage, on the other hand, is essential to prevent and control surficial erosion. Surface-drainage control, basically, consists of intercepting, diverting, slowing, and dissipating the energy of surface runoff.

Techniques or measures to achieve these objectives include:

- Interceptor drains and brow ditches
- Terrace drains
- Down (or slope) drains
- Toe drains
- Dispersal structures (level spreaders)
- Diversion dikes (berms) at top of fills or fill slopes

Guidelines for the design and construction of these measures are presented by Cedergren, 1997; Forester, 2001; Gray and Sotir, 1996; and NCHRP, 2005.

An example of an above-ground drain system to control surface water in hillside land development projects is shown on a grading plan for a typical cut slope in Figure 6.10.

6.4.2.3 Unstable Slopes and Landslides: Unstable slopes and landslide masses pose a serious challenge to hillside grading. Both cutting and filling can trigger mass slope movements by either removing lateral support or overloading a slope. Fortunately, grading can also be used to mitigate slope instability by unloading the top of an unstable slope and restraining the base with a drained toe buttress.

An existing large landslide body that cannot be adequately mitigated by loading and unloading can be further stabilized with the construction of one (depending on its size) or more shear keys that partition the slide mass into segments and intercept the slide plane by building buttresses into its body. With proper subdrainage control of the existing and anticipated water incursion (from long term irrigation), this has been a proven stabilization technique.

Adverse bedding-plane orientation, namely bedding planes that dip downslope or, worse yet, that dip into cuts and excavations can greatly exacerbate slope-stability problems and so, too, can stratigraphic sequences, where pervious, groundwater-conducting layers overlie relatively impermeable, clayey horizons. A grading plan or design should take notice of and try to avoid potential conflicts or difficulties that may arise with either of these conditions.

Examples of techniques that can be employed to mitigate or prevent landslides include the following:

- *Total removal of disturbed bedrock or slide mass.* This option is only feasible in the case of relatively small volumes relative to the size of the project.
- *Unloading the top of a slide mass* (see Figure 6.11*a*). This action may create a bench that could be used for a roadway.
- *Loading the base of a slide* (see Figure 6.11*b*). This increases the resisting moment and/or intercepts the sliding surface at the toe. A drained

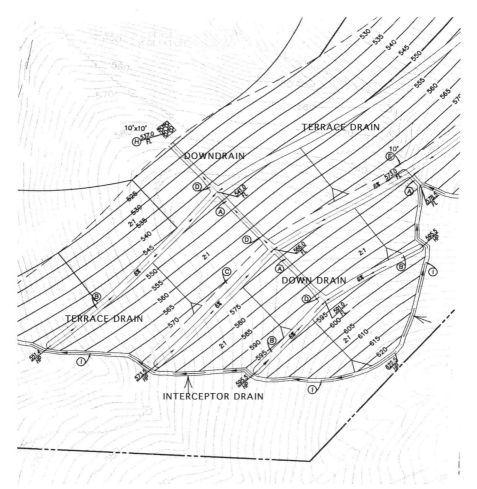

Figure 6.10 Example of above-ground drain system, consisting of brow-, interceptor-, terrace-, and down-drains, to control surface water in a hillside development.

earth or rock buttress can be used for this purpose. If earth is employed, it is important to place a drainage collection gallery between the earthen buttress and the slope.

- *Constructing a shear key* (see Figure 6.11c). A shear key is a graded, compacted earth prism that crosses the slide body and intercepts the failure surface, thus increasing shear resistance in the slope.

The installation of subsurface drainage measures are essential to assure long-term performance and are normally placed in conjunction with all these slope modification measures.

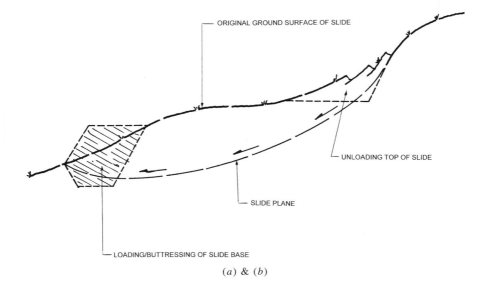

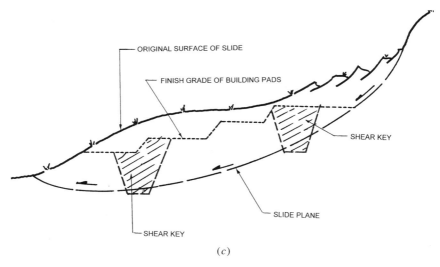

Figure 6.11 Different ways of increasing the stability of unstable slope masses: (*a*) unloading the top of a slide mass, (*b*) restraining base with a toe buttress, and (*c*) constructing a shear key.

6.4.2.4   Faults: Faults can be classified as either active or dormant. Dormant faults are historical features that can become active again at some point in the future. Movement or offset along faults occurs less commonly than landslides; however, the presence of faults can at times cause slides to occur (see Section 2.8.2 Causes of Slope Failure). But when faults are present on a property, they require special attention and precautions. In general, fault zones should be avoided and development with proper setbacks restricted in these areas. Fault zones are normally zones of highly sheared and relatively weak rock. Fault offset and previous movement can lead to the development of complex groundwater conditions in this zone. Lastly, active faults can be the focus of intense ground shaking in the event of future seismic events.

The best mitigation is to establish safe set-back zones for any development. To determine an adequacy of the offset criteria, it is critical to identify the precise location of the fault line or fault zone through extensive trenching and observation of the exposed bedrock, soil stratigraphy, offset fence lines, and other tell-tale signs of faulting activity and deformation. Along major fault zones, the stresses created in the ground are often exhibited beyond the primary fault line in the form of secondary zones of "filaments" or "splays" that have to be taken into account as well. Some states have adopted earthquake zoning laws that regulate construction near faults.

The *Alquist-Priolo Earthquake Fault Zoning Act* was passed in California in 1972 to mitigate the hazard of surface faulting to structures for human occupancy. The Act's main purpose is to prevent the construction of buildings used for human occupancy on the surface trace of active faults. The Act only addresses the hazard of surface fault rupture and is not directed toward other earthquake hazards.

In areas where development is unavoidable or where infrastructure abuts faults, overexcavation of the fault zone and placement of sand blanket cushions have been employed.

6.4.2.5   Volume Changes: Volume change and instability arise from several causes. Some soils have poor volumetric stability, that is, they are prone to shrink or swell with changes in moisture content. The so-called adobe soils, with high contents of swelling clay minerals such as sodium montmorillonite, exhibit high shrink-swell potentials. When these soils are in a dry state and are then exposed to water, they can expand with a considerable force; certainly sufficient to heave a slab foundation resting on the surface. Conversely, when these same soils are in a wet condition, they can dry out and shrink, causing foundations placed upon them to settle. Mitigation in the case of these soils usually consists of controlling water-content changes, surcharging to resist expansive pressures, treatment with lime, and use of special foundations that resist differential movement of the subgrade (Nelson and Miller, 1997).

The other source of volume change, which is equally important to grading practice, is the phenomenon of bulking and heaving. Bulking refers to the tendency of excavated soil or rock to occupy more volume, even when re-compacted, after excavation. Heaving, on the other hand, is the tendency of soil or rock to expand in situ when overburden is removed.

Different types of volumetric changes associated with earthwork and grading operations can be classified according to their origin and tendency to expand (increase) or shrink (decrease), as shown in Table 6.1.

6.4.2.6 Hard, Well-Indurated Rock: The presence of rock, particularly if hard enough that it cannot be excavated and handled by conventional grading equipment, adds an entirely new operational, logistical, and financial element to any grading operation.

The following handling and operational alternatives are generally considered:

- Ripping rock with a dozer with a single-shank attachment to achieve easier breaking and penetration. Rock hardness (up to 4,000+/− ft/sec) and fracture characteristic, and the willingness of the contractor to wear out the grading equipment, place limitations on how much can be done

TABLE 6.1   Types of Volume Change Encountered in Earthwork Construction and Grading Operations

Increase in Volume	Decrease in Volume
Bulking	Shrinkage
The denser the bedrock that is cut, the more likely the excavated material will increase in volume even if recompacted to 90 percent RC. Blasted rock can generate the most volume-change increase. Typically, in sedimentary material volume changes can range from 5 to 15 percent; in metamorphic and igneous material, 20 percent or greater	When excavated material, such as shale, clay, or silty, is placed and compacted at high-moisture contents, it will tend to dry out and shrink. Soils or shales containing expansive clay minerals are the most susceptible to shrinkage. Increasing the compactible effort and placement density minimizes later shrinkage.
Heaving	Subsidence
The tendency of bedrock to rebound or rise once overlying material (overburden) is removed from the top. This occurs particularly with dense bedrock and with deep cuts or excavations, say at depths of 100 feet or greater. Highway cuts are prone to this phenomenon, which can result in differential uplift of the road section. This action is often difficult to cure and can cause an ongoing maintenance problem until the bedrock movement reaches equilibrium.	Consolidation and compaction are both a process of densification due to the expulsion of water and air, respectively, from the voids when a soil is subjected to an external stress. Consolidation is a time-dependent process that may take months to achieve. Both result in subsidence or bulk-volume decrease. When excavated material is placed on top of a less consolidated bedrock or soil horizon for fill purposes, the weight of this overlying fill may cause the underlying layer to decrease in volume or subside.

by a dozer, even by the largest (i.e., D11), but it is by far the least expensive technique.

- Blasting is necessitated when rock hardness exceeds 5,000 ft/sec more or less, depending on the massiveness of the stone.

- "Windrowing" is placement of oversized rocks and boulders in linear mounds in deeper fills [see Figure 6.12a]. Care must be taken that soils are placed and compacted tightly around boulders and flooded so as not to allow residual voids and later localized settlements. The depth of the boulders below final grade must be sufficient so it will not create future interference with foundation and utility installation.

- Mechanically breaking with hoe rams, or other equipment, oversized material from blasting into rocks of 2 to 3 feet in size. This material can be used as a rock blanket at the base of earthen fills as shown in Figure 6.12b.

- Crushing to use as gravel or Class II subgrade base.

- Hauling and placing crushed material. These last two operations mean double handling for a contractor and will consequently increase cost significantly, threefold or more. However these costs can be recovered if the material is suitable as aggregate base for future road construction.

6.5 CUTS AND FILLS

Cutting and filling constitute the main activities of earthmoving and grading operations. Each of these operations have distinct requirements and pose separate challenges. Cuts and fills typically have different equilibrium- or stable-slope gradients. Cuts can normally be constructed at a much steeper inclination than fills because a cut normally consists of relatively undisturbed, in situ rock or soil. The steepness or inclination of slopes can be expressed in terms of (1) slope ratios, (2) percent, or (3) degrees. The relationship between these for various slope inclinations is shown in Table 6.2. Ranges of commonly used cut-and-fill slope ratios appropriate for various soil and rock types are presented in Table 6.3. A maximum slope ratio of 2:1 (H:V), whether in cut or fill, is the normally accepted standard in hillside development and mining reclamation work.

Method of placement and compaction greatly affect the stability of fills but are not a factor in the case of cuts. An exception is the construction of a cut slope that has to be stabilized by means of a buttress placed against the cut face, which in effect transforms it into a fill slope. Elapsed time since excavation or placement is a critical consideration, particularly in the case of clay soils (Abramson et al., 2003). End-of-construction is the most critical time

(*a*)

(*b*)

Figure 6.12   Incorporation of rocks at the base of fill; (*a*) Windrowing technique: fill is placed around and on top of large rocks, that are buried deep in the fill. (*b*) processed rock placed at the base of earthen fill as a rock blanket.

TABLE 6.2   Alternative Slope Inclination Designations

Slope Ratio (H:1)	Slope Percent	Slope Angle (degrees)
1	100.00	45.00
1.5	66.67	33.69
2	50.00	26.56
2.5	40.00	21.80
3	33.33	18.43
3.5	28.57	15.95
4	25.00	14.04
4.5	22.22	12.53
5	20.00	11.31
5.5	18.18	10.30
6	16.67	9.46
6.5	15.38	8.75
7	14.29	8.13
10	10.00	5.71
15	6.67	3.81
20	5.00	2.86

for the stability of fills (because of residual excess pore water pressures) whereas long-term stability, many years after construction, is a more critical consideration in the case of cuts (because of gradual loss of shear strength over time).

Many building pads and roadways are supported partially on cut and partially on fill. This juxtaposition can lead to differential settlement problems, because each zone exhibits different compressibility and settlement charac-

TABLE 6.3   Common Stable Slope Ratios for Varying Soil and Rock Conditions

Soil/Rock Conditions	Slope Ratio (Horiz.:Vert.)
Most rock	¼:1 to ½:1
Very well cemented soils	¼:1 to ½:1
Most in-place soils	¾:1 to 1:1
Very fractured rock	1:1 to 1½:1
Loose, coarse granular soils	1½:1
Heavy clay soils	2:1 to 3:1
Soft clay rich zones or wet seepage areas	2:1 to 3:1
Fills of most soils	1½:1 to 2:1
Fills of hard, angular rock	1⅓:1
Low cuts and fills (>2–3 m high)	2:1 or flatter (for revegetation)

teristics. Placing a building pad or roadway completely on a cut is not necessarily the answer to this problem. Large, high cuts can incur stability problems of their own, particularly if unfavorable geologic and hydrologic conditions are exacerbated. Large cuts also can create potential overburden handling, hauling, and disposal problems that a balanced cut-and-fill grading approach tends to minimize. Some typical cut-slope design options are illustrated in Figure 6.13.

6.5.1   Cuts and Cut Slopes

6.5.1.1   Cut Construction: When cuts are made to create building pads and sites, it is not always advantageous to build on the resulting pad directly. The presence of rock, highly expansive material, the proximity of steep bedding planes to adjacent steep, natural slopes, all warrant a different approach. In these cases, overexcavation may be required, with placement and recompaction of the native soil to a prescribed depth, to create a homogeneous building layer suitable for foundations.

When cut and fill zones meet or join along what is termed the "daylight line," they create zones of different densities and compressibilities. In such cases, the neighboring cut and fill areas are overexcavated and recompacted (see Figure 6.14) to provide underlying zones of equal characteristics and prevent or minimize long-term differential settlement between the two.

6.5.1.2   Selective Grading: In cases where certain soil types are required for fill construction that are out of sequence in the normal grading operation these soils can be mined and stockpiled for later usage. Applications would include 1) capping slopes with soils conducive to plant growth where the bulk of the cut material is sterile or has a high alkalinity or salinity or 2) capping building pads to avoid expansive material near future foundation grades. Selective grading can also be used in successive, alternating layers of permeable and non-permeable soils in fills to facilitate drainage and removal of water towards the free slope face and to prevent water accumulation within the body of the fill.

This procedure is not dissimilar to the removal, stockpiling, and later placement on reclaimed lands of top soil in surface-mining operations. This latter procedure involves double handling and can more than double the cost compared to normal grading. This additional cost, however, would tend to offset other more expensive, and often less successful, mitigation measures

6.5.1.3   Cut Slope Construction and Remediation: As cut slope are normally in well-consolidated bedrock, they frequently do not have an issue with achieving adequate compaction on the cut slope face, as in the case of fill slopes, unless the bedrock has some surficial or structural impairments. In

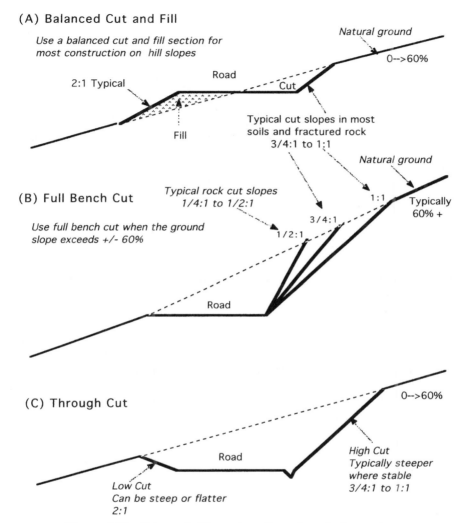

(A) Balanced Cut and Fill

Use a balanced cut and fill section for most construction on hill slopes

Natural ground

0-->60%

Road

2:1 Typical

Cut

Fill

Typical cut slopes in most soils and fractured rock
3/4:1 to 1:1

(B) Full Bench Cut

Typical rock cut slopes
1/4:1 to 1/2:1

Use full bench cut when the ground slope exceeds +/- 60%

Natural ground

1:1   Typically 60% +

3/4:1

1/2:1

Road

(C) Through Cut

0-->60%

Road

Low Cut
Can be steep or flatter
2:1

High Cut
Typically steeper
where stable
3/4:1 to 1:1

Figure 6.13   Schematic illustration of cut-slope design options.

such cases, the following remediation measures for slope-face remediation would apply:

A. *Stabilization Fills:* Sloping earthen or some times rock fill is placed against base of cut, as shown in Figure 6.15. Used primarily to prevent ravelling and minor rock falls from cut. Not performed for structural stability but, instead, more to minimize long-term maintenance problems with cut-slope face. Thickness is typically one grading-equipment width.

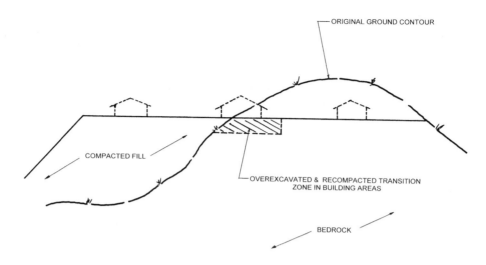

Figure 6.14 Transition zone of recompacted earth between cut and fill to prevent differential settlement.

B. *Buttress Fills:* These are structural fills to mitigate for adverse bedding conditions in exposed cuts and provide permanent slope stability, as shown in Figure 6.16. The back cut, also known as false cut, is made after the size, width, height, and the key dimensions of the buttress across the slope are determined so as to provide adequate space for it. The buttress typically has a key of several feet in depth, so it is an-

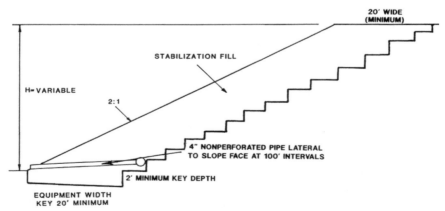

Figure 6.15 Cut-slope remediation with a stabilization fill, by placing and compacting a nonstructural protective face fill in successive lifts against the cut.

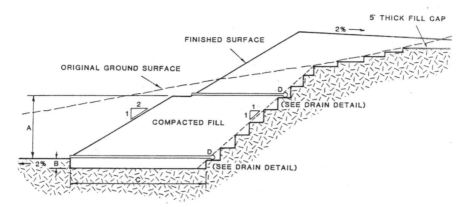

Figure 6.16 Buttress fill used to support cut and mitigate against adverse bedding conditions in exposed cuts.

chored into the bedrock below. Where stability considerations demand, the entire false cut is not made at one time but rather segmentally. Segmental construction will increase cost because of the confined space in which large equipment has to work, thereby slowing the process. It could, however, avoid a costly back-cut failure during construction, which in turn would be even more expensive to remove and reconstruct.

C. *Perimeter-Pad Cut Stabilization:* Buildable pads are sometimes created by a "daylight," or all-around cut, that is, by removing the top of a hill or ridge. This procedure can leave a residual, natural slope below the pad that is steep. A future break off or slide failure through the level pad can be avoided by means of an adequate setback and/or creation of a stabilization fill around the perimeter of the cut's outer edge consisting of the overexcavation and recompaction of a perimeter "band" of earth (typically 20 to 30 feet wide and 6 to 10 feet in depth), which is underlain by a subdrain system to divert future landscape irrigation water away from the steep natural slope below, as illustrated in Figure 6.17.

D. *Temporary or Interim Buttress:* Where long exposure of an unstable cut slope is anticipated for other earthwork logistical reasons or where segmentation is not feasible because of the size and location of a cut slope, a series of concrete inserts on graded terraces across the slope face can be installed, as shown in Figure 6.18. They typically consist of 24 to 36–inch diameter boreholes to a depth determined by the engineering geologist that are filled with concrete only (soft pile), with steel-cage lining, or with a steel "H" beam inserted into the concrete

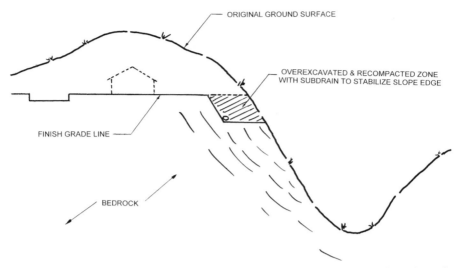

Figure 6.17 Steep slopes below pad cut protected by recompacted earthen rim buttress-and-drain system that diverts water away from natural slope below.

Figure 6.18 Temporary or interim buttress consisting of concrete, cylindrical inserts placed into a cut from graded terraces combined with a subsequent addition of "tie-backs" in the form of a steel cable installation.

(hard pile). These circular, concrete inserts are sometimes referred to as "soldier piles" or "cylinder piles." They remain in place when the permanent buttress is later constructed across the slope face.

E. *In-Place Permanent or Temporary Stabilization of Slide Mass:* Where space requirements are such that stabilization through earthwork cannot be achieved, an uphill, unstable slide mass can be "nailed" in place though the use of cable-stayed "tie backs." These tie backs consist of steel cables and concrete inserts that are drilled at an angle from the horizontal plane into the slide mass for a distance of 50 to 100 feet or more.

6.5.2 Fills

6.5.2.1 Fill Slope Construction: Fills are typically constructed at an inclination of 2:1 or flatter (see Table 6.2), depending on soil strength parameters, natural slope inclination, regulatory agency standards, or aesthetic considerations. Fills with inclinations of 2:1 (H:V) may be placed on slopes as steep as 40 to 60 percent, provided the underlying natural slope is benched and the fill is placed in compacted layers or lifts.

When steepening of a fill slope face is required due to space limitations, the fill structure should be internally reinforced. The use of geogrid or geotextile reinforcement layers can permit steepening of a fill face to 1:1½ (H:V). Fill slope design options are illustrated schematically in Figure 6.19.

6.5.2.2 Deep Fills: In previous operations, canyon and valley fills were typically and uniformly compacted to 90 percent density. Long-term experience, however, has shown that when they are deeper, they are susceptible to settlement. Consequently, more recent industry standards call for fills depth to exceed 70 feet $+/-$, to be compacted to 95 percent for the lower 50 feet.

6.5.2.3 Fill Slope Remediation: One of the problems with fill slopes can be inadequate compaction (low density) near the face because of lack of lateral confinement. This problem can be overcome by either (1) overfilling an equipment width and then cutting back to final grade or (2) building the fill normally and then following up with dozer tracking, a sheepsfoot roller and finally a grid rolling of the slope face.

Compaction to high density, especially through the use of overfilling and cutting back and a grid roll finish, generally improves engineering properties of a soil—e.g., strength and volume stability—but not its ability to support plant growth. The finished slope face product presents a smooth, extremely hardened surface; to plant larger size container material—e.g., trees and shrubs—almost requires jackhammering. The most desirable finish is the one

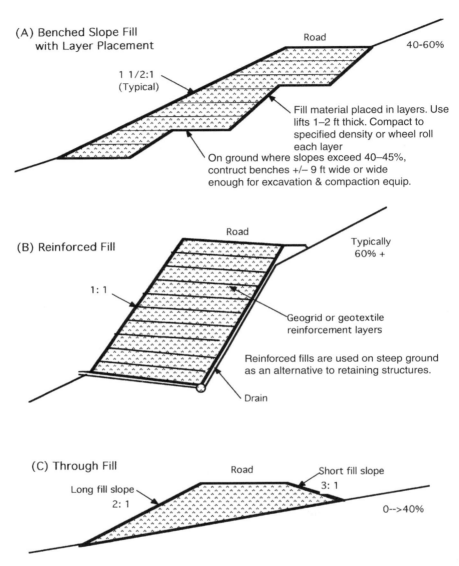

Figure 6.19 Schematic illustrations of fill-slope design options.

with a sheepsfoot roller, and while it still achieves the necessary compaction, it leaves a roughness behind that allows the seed material and stabilizer in the hydromulch to get a foothold and enables quicker germination on the slope face. Ways to balance plant growth and engineering stability needs are described in Chapter 3, Section 3.6.5.

6.6 EROSION CONTROL DURING GRADING

While a project is graded and expected to go through a rainy season or seasons, it is a standard practice and a requirement by most agencies to have the site itself and any downstream, off-site properties protected through measures that normally include interim seeding, installation of detention and desilting basins, and other methods. Erosion control principles are discussed in Chapter 2, Section 2.7. Various erosion-control measures are described in detail elsewhere (Goldman et al., 1986).

Both federal and state requirements mandate that erosion-control measures be employed during grading operations. These requirements are referred to as Stormwater Pollution Prevention Plans (SWPPP). These plans vary from state to state but typically require the following:

a. Interim groundcover protection that typically includes a quick germinating seed mix with an added emulsion tackifier to keep the soil in place until germination takes place.

b. Detention basins to reduce runoff velocities by calming water flow in the basin, retaining excess runoff, and discharging the runoff in a controlled fashion.

c. A desilting basin to capture the expected silt and sand produced from early unprotected runoff of a freshly graded site. The desilting basin allows the sediment to settle on the bottom and then discharges clear water above through an overflow pipe.

d. Often detention and desilting basins are installed as a combination facility.

Some examples of these runoff- and erosion-control measures are shown in Figures 6.20–6.23.

6.7 ECONOMICS OF GRADING

Grading has significant cost implications and often represents the single largest cost factor—be it hillside land development, highway construction, or overburden removal and replacement in surface mining.

Mine reclamation grading operations, while having certain limitations, are still not nearly as restrictive as those for hillside land development, where literally every square foot of level, buildable ground generated has extreme value. Space confinement and constraints on movement are also greater in hillside development work. Additionally, in the preparation of the grading plan, numerous other constraint must be considered and avoided, such as providing gravity flow for sewers, controlling, collecting, detaining, and dis-

Figure 6.20   Chopped, dry straw mulch applied to slope with blower.

Figure 6.21   Hydromulching application on a slope.

Figure 6.22   Detention-desilting basin used to control runoff and sediment.

Figure 6.23   Intake structure in a detention-desilting basin surrounded by gravel filter, trash rack, and antivortex plate.

TABLE 6.4   Cost Estimating Form for Various Line Items in a Hillside Grading Operation

Line Items	Unit Cost Measure
Preliminary Operations:	
Mobilization	
Demolition, clearing, and grubbing	
Prewatering	
Alluvium removal	
Colluvium removal	
Landslide removal	
Canyon subdrain, perforated	
Canyon subdrain, solid	
Mass Excavation:	
Buttress or stabilization fill, out and in	
Buttress or stabilization fill drain, perforated	
Buttress or stabilization fill drain, solid	
Shear key, out and in	
Shear key drain, perforated	
Shear key drain, solid	
"Back-cut" failure removal	
Daylight cut buttress	
Daylight buttress drain, perforated	
Daylight buttress drain, solid	
Overexcavation for transition pads	
Overexcavation for cut pads	
Key excavation at toe of fills	
Key drains	
Rock-heavy ripping	
Rock-drill and blast	
Rock-deliver to breaker or crusher	
Rock-mechanical breaking	
Rock-crush to 4 inch or Class II size	
Rock-disposal of oversized material (including windrowing and rock blankets)	
Overexcavation of rock pads	
Line-blast street sections for utilities	
Overexcavation of rock in street sections for utilities	
Selective grading	
Final Operations:	
Terrace drains	
Down-drains	
Interceptor drains	
Toe drains	
Splash walls	
Slope finish	
Lot finish	
Debris walls	
Rockfall protection	
Detention basin	
Erosion control	
Dust control	
Export	
Import	

TABLE 6.4 (*Continued*)

Line Items	Unit Cost Measure
Special Operations:	
Groundwater control	
Wet material handling, spreading, drying, and replacing	
Gravel and fabric blankets	
Surcharge fill	
Selective grading	
Endangered species; Migratory Bird Act; archaeological and paleontological conditions; rainy season constraint costs	

charging storm-drain runoff, and meeting vertical and horizontal road-alignment standards that satisfy demands of traffic safety and fire and other emergency needs. Mine reclamation has typically a vaster area to operate in and does, in most cases, not need to concern itself with residential and commercial development, road geometrics for future public use, and utility services. It still has to deal with sometimes steep topography and runoff from mined areas. Potentially contaminated runoff from these areas requires containment, treatment, and control of discharge.

To structure and plan the most efficient and cost-effective grading operation requires good earthwork logistics. In other words, the art of hauling dirt in machines has become a science in itself. Critical are such issues as minimizing the hauling distance (the longer the haul the more expensive), hauling downhill with loaded scrapers, hauling uphill empty, and so forth.

A typical hill-grading project cost estimate form depicting the various and numerous line items involved in such an operation can be used to illustrate how many complex operations have to be undertaken and how costly this can be. An example of a cost form for estimating various line items in a typical grading operation is shown in Table 6.4. In such a hillside project, where the natural topography exhibits 15 to 20 percent or higher slope gradients, grading can amount to between 25 and 35 percent of the total development costs, which represents the highest single hard construction cost item in such a budget. This is well above the individual cost for roads, sewers, water, storm drains, or dry utilities. Often the next highest line item cost is for landscaping and irrigation installations—because the more grading, the more the area is disturbed that must be landscaped—followed by storm-drain costs to manage runoff across the disturbed terrain.

The shortage of undeveloped, level land and public tolerance for high land costs in today's markets are the main factors that justify the cost of such an expensive undertaking. Furthermore, hillside housing provides other amenities that generate added revenues, which in turn justify higher grading costs. Added amenities and attractions include lower density, more open space, scenic vistas, and recreational opportunities.

6.8   REFERENCES

Abramson, L. W., T. S. Lee, S. Sharma, and G. M. Boyce. 2002. *Slope Stability and Stabilization Methods*. 2nd ed. New York: John Wiley & Sons.

Goldman, S. J., K. Jackson, and T. A. Bursztynsky. 1986. *Erosion and Sediment Control Handbook*. New York: McGraw-Hill.

Gray, D. H., and R. Sotir. 1996. *Biotechnical and Soil Bioengineering Slope Stabilization*. New York: John Wiley & Sons.

Forester, K. 2001. *Drainage for Slope Stabilization*. ASCE Press, New York, NY.

Cedergren, H. R. 1997. *Seepage, Drainage, and Flow Nets*. New York: John Wiley & Sons.

NCHRP (2005). *Environmentally Sensitive Channel- and Bank Protection Measures. NCHRP Report 544,* National Cooperative Highway Research Program, Transportation Research Board, 50 pp., Washington, DC.

Nelson, J. D., and D. J. Miller. 1997. *Expansive Soils: Problems and Practice in Foundation and Pavement Engineering*. New York: John Wiley & Sons.

7 Principles of Landform Grading

7.1 INTRODUCTION

Landform Grading is a technique that attempts to mimic stable, natural landforms. This approach has evolved specifically in response to deficiencies in traditional or conventional methods of slope grading; but it also applies to landform shaping in general. These deficiencies include a visual product of flat surfaces and rigid, rectilinear-slope forms with little resemblance to the original, natural landscape.

Landform Grading concepts were developed to redress these deficiencies and to introduce aesthetic considerations into hillside development and mining-reclamation projects. To better understand the basic principles of Landforming and to differentiate them from conventional grading practice, it is helpful to review these earlier or traditional methods and their basic characteristics.

7.2 THE TRADITIONAL METHOD

7.2.1 Conventional Slopes and Their Design Elements

7.2.1.1 Slope Plan and Profile Shape: Conventionally graded slopes are characterized primarily by rectilinear- and planar-slope surfaces with unvarying gradients and angular-slope intersections.

The slope tops exhibit a linear crestline devoid of any topographic relief. The bottom of the slope exhibits a linear and angular intersection with the base. Both these features not only encourage but also dictate the linear placement of structures or homes and road alignments. Examples of conventional grading for hillside housing developments are shown in Figures 7.1 and 7.2.

7.2.1.2 Drainage Devices: Slope-drainage devices are usually constructed in a rectilinear configuration (parallel for terrace drains, perpendicular for down-drains) and in prominent and highly exposed positions, maximizing the negative visual impact on the slope face, as shown in Figure 7.3. No consideration is given to screening these devices, which would be difficult in any event on planar or uniform surfaces.

Figure 7.1   Illustration of conventional slope depicting grading plan and profile shape.

7.2.2   Building Pads

Pads are configured uniformly, typically rectangular, and always bounded by rectilinear perimeters. This arrangement tends to produce visual monotony (see Figure 7.2).

7.2.3   Landscaping

Landscaping is applied in rigid, uniform patterns, and plant material such as trees and shrubs are placed at equal spacing to achieve the stated conventional objective of "uniform coverage," which further emphasizes the artificial, man-made environment. "Uniform Coverage" means trees are typically planted at 15-foot intervals and shrubs at 3-foot intervals, whereas seeding and hydro-mulching of the ground cover is uniformly applied to the entire surface. Typical examples of these artificial slope elements and what uniform coverage means are shown in Figures 7.4 and 7.5.

The uniform coverage approach likely arose out of the linear-thinking process discussed in Chapter 1. Under this reasoning, the goal is to provide for quick and uniform coverage of artificially created, planar-slope surfaces and for added protection against erosion and siltation. Such surfaces, aside from having little or no redeeming aesthetic features, are also devoid of creative opportunities for either the placement of drainage devices or plantings that would mimic the characteristic natural landscape of the area.

Figure 7.2 Photo of conventional grading plans for hillside development.

Occasionally, slopes are landscaped in a purely ornamental or decorative manner, as shown in Figure 7.6. This approach tends to look artificial, and it is unlikely to be compatible with slope hydrogeology.

7.2.4 Historical Use and Observations

The traditional or conventional approach is still used in most grading applications, whether it is hillside land development for housing, waste-landfill

Figure 7.3 Conventional slope grading with horizontal benches and exposed vertical down-drain.

Figure 7.4 Conventional slope grading with linear, vertical down-drain and uniform plant coverage.

Figure 7.5 Appearance of conventionally graded slope with a uniform and rectangular tree-planting pattern.

Figure 7.6 Example of purely ornamental or decorative slope planting.

design, or surface-mining reclamation. For decades planners, engineers, landscape architects, geotechnical engineers, and surveyors have been trained in this technique, and they and their clients feel most comfortable operating in familiar territory. Familiarity also provides confidence and reasonable assurance of expected expenditures. On the other hand, the unfamiliarity of new approaches may cause some reluctance and skepticism, particularly if combined with automatic expectations of higher costs whether rightly or wrongly.

7.3 THE IMPROVED METHOD

7.3.1 Contour Slopes and Their Design Elements

7.3.1.1 Slope Plan and Profile Shape: Contour graded slopes deviate from conventionally graded slopes in that the slopes are curvilinear in plan rather than linear. But, because the transitions between concave and convex elements are very gradual, they create repetitive undulations, resulting in an "ocean-wave" effect. These undulations are an attempt to achieve a pleasing visual effect rather than mimicking any natural topographical feature. The slope form and landforms are not based on natural analogs but instead are random and intended to create an aesthetic effect. The downslope gradients are still mostly unvarying, and profiles are linear (as opposed to curvilinear).

Transition zones to natural slopes and slope intersections generally have some rounding applied. On the other hand, slope profiles at the top are still linear, that is, there is little or no vertical relief or deviation from a straight edge at the top. The same applies to the toe of slope. An example of an "improved" or contour grading plan is shown in Figure 7.7.

7.3.1.2 Drainage Devices: Slope-drainage devices are usually still constructed in highly exposed positions on the slope face. Often down-drains are located in the concave curvature of the slope segment in a vain attempt to screen them from view. However, this approach tends to interfere with the establishment of a diverse vegetation regime on the slope. Furthermore, as a result of the gradual transition, there is no other room to hide or screen these drains.

7.3.2 Building Pads

Pad outlines are typically still traditionally configured in a rectangular or perhaps somewhat curvilinear fashion. Occasionally, they may display an irregular perimeter but never to the extent that they deviate much from polygonal areas with linear boundaries.

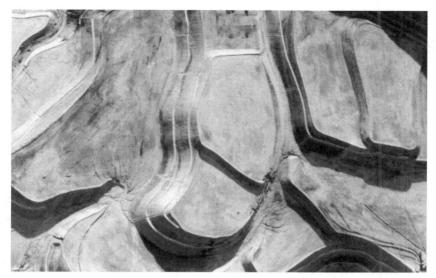

Figure 7.7 Aerial view of improved, or contour, slope grading.

7.3.3 Landscaping

Landscaping is applied very much as in the conventional approach, because uniform coverage is the dictate of the industry. Occasionally, it is installed in random or geometric patterns, but the landscaping does not display any natural plant-distribution patterns.

Attempts to "tuck" away the concrete down-drain devices in the concave, swaled portion of the slope face prevents the introduction of a natural, denser vegetation palette with trees and shrubs in this locale. The concrete devices prevent water from collecting in the swale and being absorbed by the soil thereby preventing this area from supporting and benefiting from a concentration of plant life.

7.3.4 Historical Use and General Observations

The contouring approach is most commonly employed in the design and construction of golf courses, where the intent is not to recreate a natural topography with all its elements but rather to create aesthetically pleasing land forms and a challenging play environment. Under certain conditions, whether by a requirement of the approving agency or the desire of a particular developer to enhance the marketability of the project, some slopes—in entry or other key locations—will be contour graded. The design in this case will typically originate from the desk of a landscape architect and/or an opportunistic land planner.

7.4   THE ENVIRONMENTALLY RESPONSIVE NEW TECHNIQUE

7.4.1   Landform Slopes and Their Design Elements

7.4.1.1   Slope Plan and Profile Shape: The *Landform Grading* concept seeks to replicate the irregular shapes of natural slopes. In contrast, most man-made slopes using the previous techniques have been designed and graded primarily to create elevations and profiles with an unnatural, artificial appearance. Landform graded slopes, on the other hand, are characterized, instead, by a continuous series of distinctive concave and convex forms interspersed with mounds that blend into the profiles—by nonlinearity in plan view and by varying slope gradients. The main characteristics and differences between landform graded slopes and conventionally graded or contour graded slopes are illustrated schematically in Figure 7.8.

Where room permits (and mine tailing and landfill designs are prime candidates for this approach) and where compatible with regional topography, the profile of slopes (particularly in the case of fill slopes) exhibits a concave form with steeper gradients near the top and with gradually decreasing, flatter inclinations near the bottom. As explained in Chapter 4, a concave-slope profile is more stable than a linear profile, and it more closely resembles the equilibrium profile of natural slopes. Additionally, there are significant transition zones between man-made and natural slopes to avoid abrupt changes in landforms and to create a minimally perceptible blending of the two, as the terrain moves from man-made to natural.

7.4.1.2   Drainage Devices: The hydrologic flow regime in landformed slopes differs significantly from that in conventionally graded slopes. The surface runoff is directed radially into swales as opposed to being dispersed evenly over the slope in a linear, sheetflow pattern, as shown in Figure 7.9.

This concentration of flow may seem counterintuitive, but the landform-revegetation plan is designed to accommodate and to use the extra water that is delivered to the swale areas. Where soils (fill slopes) or bedrock (cut slopes) conditions permit and geotechnical stability analyses support the approach, all terracing and concrete devices can either be eliminated or reduced in quantity and frequency.

Where tributary-runoff discharge is greater than acceptable and the soils more erosive, swale sections can be reinforced with geofabrics—e.g. erosion-control blankets (ECBs) or turf-reinforcement mats (TRMs). This will still allow for the planting of trees and shrubs in the general swaled areas as well as the absorption of water into the soil while simultaneously stabilizing the swale during the initial stages. Once established, vegetation will provide better soil stabilization and erosion control.

Where hardened slope-drain devices are required by government regulations, through the recommendation of the geotechnical engineer or because of excessive tributary drainage, these devices can be placed in a variety of

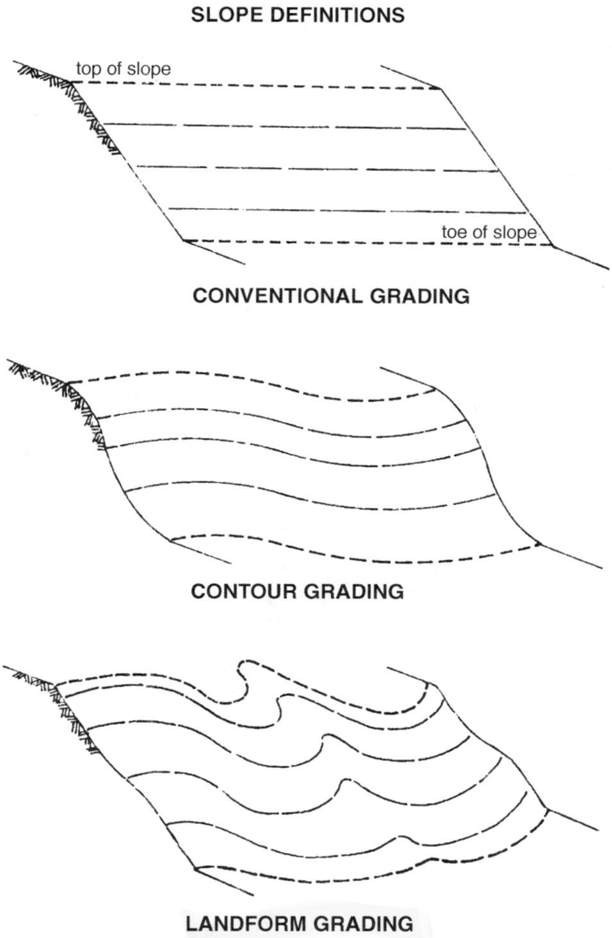

SLOPE DEFINITIONS

CONVENTIONAL GRADING

CONTOUR GRADING

LANDFORM GRADING

Figure 7.8   Comparison of plan views and profile shapes of landform graded slopes versus conventionally graded and contour graded slopes.

alternate ways. The drains can either follow natural-drop lines in the slopes, or they can be tucked away between a concave swale and convex berm combination to conceal the drains from view. These alternatives are accomplished in the same manner as in the case of a natural-slope analog. The drainage swales are placed in either a diagonal, curvilinear, or elbow-shaped manner across the slope face. A major advantage of these alternative configurations is that channel length is increased, resulting in the reduction of gradient and flow velocity in the drain.

To enhance the appearance of exposed segments of concrete devices in high visibility areas, they can additionally be treated with natural river rock

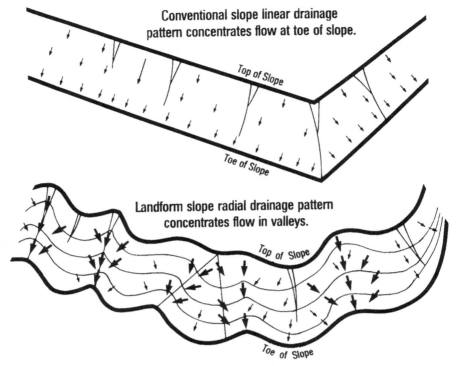

Figure 7.9 Comparison of surface water-flow regime in a landform graded slope as opposed to a conventionally graded slope.

or cobblestone. This not only gives the drainage device a natural "stream bed" appearance but also provides a secondary benefit of increasing friction and thereby reducing flow velocity in the drainage channel. The appearance and shape of "hardened"-downslope drains in a conventional- versus landform grading application are compared in Figure 7.10.

7.4.2 Plateaus and Building Pads

Plateaus created from removing or cutting off hilltops and the building pads created for urban development do not have rectangular shapes or linear perimeters; but they do follow, unless prevented by overriding conditions, either irregular shapes that fit into the terrain or that follow underlying contour lines, as shown in Figure 7.11.

(*a*)   (*b*)

Figure 7.10   Photos of drainage devices on graded slopes: (*a*) conventional and (*b*) landformed.

7.4.3   Revegetation Landscaping

Unlike other approaches, landscaping in *Landform Grading and Revegetation* is not regarded as a regimented, uniform plant application process. Instead, landscaping is motivated by the goal of replicating the original vegetation palette or distribution of tree and shrub types that existed on-site before ground disturbance and grading. In other words, trees and shrubs are selected, preferably from the original plant community, and placed in patterns characteristic of natural plant patterns in the vicinity.

In this approach, trees and shrubs are concentrated largely in the concave slope areas and swales, while convex or interfluve areas are planted mainly with grasses and other ground covers. This distribution of woody and herbaceous plants occurs naturally in nature: Trees tend to concentrate in wetter swale or valley areas, while grasses grow in drier convex or interfluve areas, as shown in Figure 7.12.

Spacing under landform revegetation also attempts to replicate natural conditions. Trees and shrubs are clustered in groups as opposed to being spread out uniformly on a slope as shown in Figures 7.13 and 7.14. Needless to say,

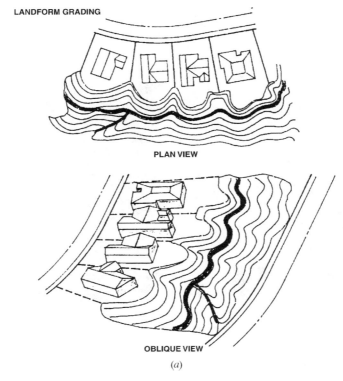

(b)

Figure 7.11   Landform graded hillside development: (a) schematic drawing showing irregularly shaped building pads in addition to curvilinear slope shapes and (b) aerial photo of landform graded hillside development.

Figure 7.12   Distribution of woody vegetation (trees and shrubs) and grasses in a natural hillside. Woody plants tend to cluster in swales while grasses grow in drier interfluves.

Figure 7.13   Example of landform grading and revegetation in a hillside housing development *during* initial planting stage. Trees and shrubs are clustered in hillside swales.

Figure 7.14 Example of landform grading and revegetation in a hillside housing development *after* plant establishment (mature stage). Trees and shrubs are clustered in hillside swales and grasses in convex-slope areas.

this "revegetation" concept and process requires the presence of underlying "landform"-shaped slope forms.

7.5   REPAIR AND REJUVENATION TECHNIQUES FOR EITHER MAN-MADE OR DAMAGED NATURAL LANDSCAPES

7.5.1   General Observations

Landform Grading or *Landforming* techniques are applicable for both new construction in virgin or undisturbed terrain and for repair or rejuvenation of existing landscapes. With regard to the latter goal, landforming provides interesting possibilities and alternatives to traditional approaches to erosion and sedimentation control, with their primary emphasis on structures and erosion-control products. Landforming is not limited to grading operations in hilly terrain with high relief, but it is also a way to add topographic interest to relatively flat or level terrain. Landforming and revegetation can be viewed as a creative tool or approach for landscape architects in which the underlying slope and landforms play an integral role in the landscape plan.

7.5.2   Direct Slope Replication

When terrain with high relief is impacted by regrading for land development, highway construction, mining, or other alterations, large portions of the newly regraded area will consist of a residual slope component. It would be ideal to replicate and incorporate the original underlying slope forms into the newly manufactured slopes to return those prominent features to a semblance of their natural state. Constraints on this approach are probably greater for rejuvenation and repair of conventionally graded as opposed to de novo construction, but it is nevertheless worthy of consideration.

A conscious attempt was made to replicate the original underlying slope and valley topography in the valley fills in the Hollywood Hills, California, mass-grading project. Both diagonal and elbow-shaped concave features were incorporated into the fills. The surface of these fills now occurs at a higher average elevation but still closely resembles the original topographic surface in its most important respects. The Hollywood Hills mass-grading project is described and discussed in detail in Chapter 12.

Another example of the potential for the "direct-slope replication" approach exists in its use on disposal tailings from surface coal mining operations in Appalachia for valley fills. Instead of using the conventional dam configuration to terminate a valley fill in Virginia (see Figure 7.15), the fill could have been placed in a way to replicate adjacent natural valley slopes but at a slightly higher average elevation. Nature provides the template for direct-slope replication in the vicinity, waiting to be recognized. This "replication" approach is preferable to manufacturing an alien landform that is out of character with the entire surrounding landscape.

7.5.3   Complete or Partial In Situ Landform Restoration

During an inspection of an area in the process of reclamation in Virginia, a natural hill was observed projecting above an otherwise level ridge. Upon closer examination, a power pole was observed that was firmly planted on the peak of this hill (see Figure 7.16). The hill broke up a monotonous, manufactured ridge line and created a more natural ridge-line effect that was common to the area. The mine operator advised that the only reason this hilltop was retained is that the power provider would not allow this pole to be moved. So it was not by design but out of a secondary necessity that an appropriate landform element was created at this location. One might ask why do we not assume more "imaginary power poles" strategically placed to help us perceive and construct more natural ridgelike landforms?

Another example of an "accidental" landform preservation occurred in the case of a high wall created by mining in New Mexico. Under state and federal regulations, artificial high walls must be cut back to safe angles, typically at $3:1$ (H:V). Because of delays in implementing this work, natural vegetation began to establish itself at the foot of the bluff, including an endangered species of a rock rose (see Figure 7.17). A battle between mine regulators and endangered species regulators ensued about who had primary jurisdiction. Hopefully, the status quo has prevailed at this site for the sake of preserving both a naturally appearing, artificial bluff and the endangered rock rose.

(a)

(b)

Figure 7.15 Tailings-disposal fill configurations: (a) "outward"-facing slope of valley fill once constructed and graded and (b) adjacent natural slope, whose topographical form would have made an ideal natural analog for the "outward" slope.

Figure 7.16 Photo of unintentional ridge-line variation that provides considerable visual interest and relief to the monotony of linear-ridge profiles.

Figure 7.17 High wall or bluff left over from surface mining with rock rose at foot.

On a much larger scale, the mountain-removal and valley-fill process practiced in the Appalachian Mountains has caused radical alterations of the landforms and vegetative cover. This alteration, which is associated with coal-mining operations, includes the filling of miles of valleys and streams with artificial fills as high as 1,000 feet. Even though regulations mandate that reclamation has to be performed in such a manner that the topography will be restored to an "approximate original contour" (AOC), the actual results tend to be quite different. When questioned about this discrepancy, a mining industry representative explained that "AOC resides in the mind of the beholder." Apparently, this concept is allowed by regulatory agencies; not surprisingly it has also been subject to differing interpretations.

A design that would at least mimic the original contours would be one that recreates a three-dimensional landscape of mountaintops or hilltops and valley forms. Drainage in these valley floors could be handled much more efficiently with meandering streams reinforced with durable rock lining. This would allow a more natural concentration of woody vegetation—i.e., trees and shrubs—along the valley bottoms. Instead, under the present scheme, the runoff or drainage is dropped straight down a 2:1 out-slope over great heights with no opportunity for enhancement or diversity of vegetation. A conventionally placed valley fill generated from the mountaintop removal process is compared in Figure 7.18 with a proposed alternative that replicates or mimics the original topography.

There are certainly good reasons for a limited amount of level ground in mountainous terrain for such developments as housing, schools, shopping centers, golf courses, prisons, sports fields, and so forth. The price to be paid for creating this level ground and associated land uses should not come, however, at the expense of damage to the beauty and environmental integrity of vast tracts of the original landscape that have attracted generations of residents and tourists alike. Both mining and land reclamation and restoration can be achieved in this region in reasonable harmony with one another by applying *Landforming* principles and techniques.

7.5.4   Creation of New Physiographic Landforms

Some landform alterations occur on level or near-level terrain, and the underlying topography does not provide for local examples to replicate. In such a case, a new topography must be created that permits harmonious blending as it provides for long-term stability and allows the topography to continue to mature and to develop into a self-sustaining landscape that provides for local vegetation and animal habitat. This is particularly applicable in such cases as the oil-sands operations in northern Canada (see Chapter 11) and landfills in level terrain, that is, in southern California.

7.5.5   Slope-Form Restoration via Landform Grading

Another option for slope-form restoration is where the underlying topography does not provide adequate examples. In this case, the standard forms identified

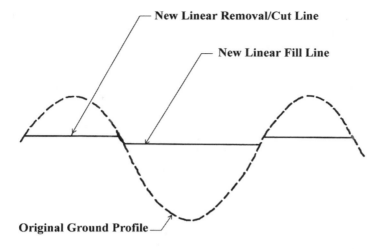

New Linear Removal/Cut Line

New Linear Fill Line

Original Ground Profile

FREQUENT CURRENT APPROACH - NOT A.O.C.

Mountains and Valleys flattened - No Semblance of Original Contours

(*a*)

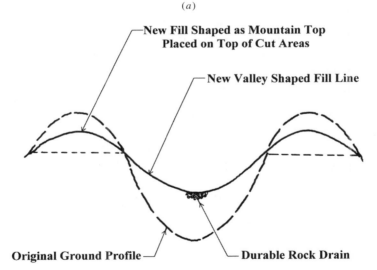

New Fill Shaped as Mountain Top Placed on Top of Cut Areas

New Valley Shaped Fill Line

Original Ground Profile   **Durable Rock Drain**

PROPOSED ALTERNATIVE - A.O.C. ACHIEVED
Mountains and Valleys in Form Restored

(*b*)

Figure 7.18 Alternative cross sections of valley fills generated from mountaintop removals: (*a*) Valley fill not consistent with approximate original contour (AOC). (*b*) Valley fill that approximates original topography.

in Chapter 8 should provide some guidance. A whole suite of landforms—
e.g., swales, elbows, wishbones, cones, and other slope shapes—are identified
in Chapter 8 that could work effectively in different physiographic settings.

7.5.6   Landform Restoration after Mass Grading and Filling

Massive alterations of landforms can occur through natural processes such as
watershed erosion (see Asaayi Lake case study, Chapter 11, Section 11.3),
through mountaintop removal and valley fills in Appalachian Mountain coal-
mining operations (see Section 7.5.3), or through mass grading and canyon
filling (see Hollywood Hills case study, Chapter 12, Section 12.1). In these
cases, landform grading provides a way of restoring a semblance of the original topography and appearance of the natural terrain.

When visiting a surface coal-mining operation in Virginia at the invitation
of the Commonwealth of Virginia, Department of Mines, Minerals, and Energy, the senior author inspected an ongoing mine operation. Several overburden spoil dumps were observed that resulted from mountaintop-removal
operations intended to reach underlying coal seams. Two pronounced, flat-
topped dumps drew special attention. According to a mine representative,
these disposal sites were considered complete in their current form. The Division of Mine Land Reclamation requested a recommendation on how landform grading would alter this approach to spoil disposal.

No flat-topped plateaus of natural origin were visible in the area, instead
the surrounding terrain was characterized by ridges, hilltops, and valleys. This
was the topographic template for the region. This template would also be
consistent with the Federal Guidelines (e.g., AOC), which specified replicating the original topography as closely as possible. A landforming approach
would satisfy this requirement. As a further bonus, this approach would provide additional room for overburden storage in close proximity to existing
mining operations (avoiding a proposed long haul to another valley fill) and
make it unnecessary to destroy other canyons with fills.

This would have more than offset the additional cost of "stacking" the fill
on the plateau to a peak formation.

The mine provided an "as-graded" topographic map of the existing conditions as shown in Figure 7.19. To demonstrate the change, a new grading
plan was developed as an overlay, showing how additional fill could be placed
on the current plateaus to recreate a hilly topography, as shown in Figure
7.18.

Ultimately, the mine agreed that the landform grading approach would be
beneficial and advantageous. This conclusion on their part, unfortunately, was
based strictly on their perception that adopting this new grading approach
would allow them to perform even more mountaintop removals. They reached
this conclusion because they believed that this new approach would permit
them to place the surplus spoil from additional mountaintop removal in an

Figure 7.19 Topographic map of existing site conditions at overburden spoil–disposal mound.

pristine valley—the very outcome that this new grading approach was designed to prevent!

7.6   SURFACE MINING RECLAMATION

7.6.1   Impact of Surface Mining

Nothing has a greater impact on natural landforms than the large-scale surface mining that is conducted today with massive earthmoving machinery. The level and scale of this surface mining can alter in a few days what took natural geomorphic forces hundreds of thousands or even millions of years to form and shape.

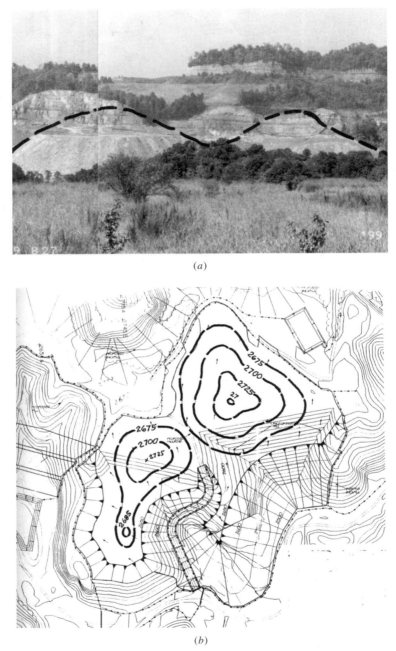

(a)

(b)

Figure 7.20 Proposed new Landform grading plan for spoil disposal showing impact of revised grading: (a) photo overlay and (b) grading plan.

In the case of geomorphic impacts associated with hillside housing developments in the western United States, this alteration largely consists of taking natural hill and valley topography and reconfiguring it into building pads either on flat plateaus (former hilltops) or on a system of linear hillside terraces and benches. In the case of surface extraction of coal in the Appalachian region of the eastern United States, on the other hand, this alteration reduces mountain and valley terrain into gigantic, flat plateaus, for which (in most cases) a purpose has yet to be found. In contrast, the oil-sand extraction mining in Northern Canada begins with a flat landscape and transforms it into a man-made, three-dimensional landscape of hills, dumps, and pits or lakes.

The construction of waste-disposal landfills in relatively flat geographic regions involves a process of building artificial mounds surrounded by a containment envelope consisting of layers of compacted soil, drainage media, and impermeable membranes. Some of these landfills can reach surprising heights as a result of a practice known as "vertical expansion" (Qian, Koerner, and Gray, 2002). Areas comprised of old sand and gravel pits or rock quarries are also characterized by terrain that has been significantly altered, both in terms of its geography and hydrology. It is ironic that, where three-dimensional relief existed, it has often been converted into flat, two-dimensional terrain. Conversely, where a flat two-dimensional landscape previously existed, that landscape has now been transformed into a quasi-three-dimensional system.

The language used to describe some of these geomorphic alterations and conversions is somewhat revealing. For example: a tailings deposit is referred to on engineering drawings and in day-to-day vernacular as a "DUMP" or "SPOIL PILE." When terminology such as this is used, one tends to get the corresponding results. If a topographic feature is termed a *dump*, it is designed as a *dump*, it is surveyed as a *dump*, and finally it is built as a *dump*. Under these conditions, it is not surprising if it looks like a dump as well. A dump typically has no natural features, no redeeming environmental or aesthetical qualities; it may not even exhibit long-term stability.

In a flat landscape, the introduction of a 100-foot high, 3-mile long, and 1-mile wide spoil pile of "dirt" or mine tailings, with a planar face and unvarying slope gradients, is more than a dump. It also represents a de facto landform, albeit an artificial one, that geomorphic agents (namely, wind, water, and gravity) will reform and reshape over time. This reshaping and reforming process was discussed and described in detail in Chapter 5 (Section 5.7.1, Evolution and Morphometry of Spoil Mounds). Also Figure 5.2 shows the profiles of spoil mounds that were reshaped by erosion over time. Similarly, the application of digital terrain models, described in Chapter 5, Section 5.8, make it possible to estimate how anthropomorphic landforms—for example, tailings mounds—will evolve and change over time (see Figure 5.12). This modeling permits estimates of surface lowering, soil loss, and gully incision at different elapsed times.

So, aside from the immediate visual impact and intrusion of features such as spoil mounds, long-term maintenance will be an area of concern as well, because conventional design of these man-made features results in a landform that may be subject to higher rates of erosion and lower long-term stability compared to natural landforms.

7.6.2   Importance of Replicating Original Topography and Hydrology

What happens when coal mining in southwestern United States occurs in a landscape characterized by mesas and washes and if this mining activity reshapes the original landscape into a terrain more closely resembling rolling Iowa farm land? It basically means that no consideration has been given to implications of the original topography and equally none to the underlying, original hydrology. The importance of studying and preserving the original drainage networks and drainage densities was discussed in detail in Chapter 4, Section 4.6. As pointed out earlier, natural watersheds tend to have drainage networks and slopes that are relatively stable with respect to one another. Forces associated with rates of runoff and hillside gradients, which cause channels (or gullies) to develop, tend to be in dynamic equilibrium with forces that resist erosion, such as shear strength or allowable tractive stress. Under these conditions, deep incisions or channel erosion is minimized and sediment yield comes primarily from rainfall erosion processes, such as raindrop splash, sheet, and rill erosion.

The main concern of engineers in altered landscapes, following surface mining, is "control" of drainage. Unfortunately, this control is too often promulgated without sufficient attention to the preexisting natural hydrology. Natural drainage networks evolved over long periods of time and are a reflection of the underlying soil and bedrock conditions and the form and size of tributary drainage areas. These conditions and circumstances, in turn, provide the basis for the establishment of vegetation with regard to type, growth, and distribution patterns.

As noted in Chapter 4, studies of postmining-reclamation landforms often reveal a different picture. Inspection of reclamation sites in the midwest showed, for example, that even when graded to acceptable specifications, freshly graded surfaces may be left with a high potential for advanced erosion, including channel incision and gullying. These erosion processes, in turn, can disrupt attempts at revegetation and produce sediment that must be contained in sediment ponds or behind check dams to prevent off-site damage. These postmining-reclamation sites were characterized by drainage networks consisting of rock lined channels and slope profiles that differed substantially from those of stable, natural watersheds or land surfaces.

A similar problem existed in the former East Germany, where the Soviet Union extracted coal and uranium from below the surface of rolling, hilly terrain. Under the former communist system, there was little if any reclamation or mitigation of mining damage of any kind; so, open pits that extended for kilometers were left behind. After unification, the German government began a process of rehabilitation, which consisted primarily of

"filling in the trenches and holes." This basically resulted in a slightly inclined planar surface crossed by rock-lined drainage channels. These channels
more closely resembled a plumbing system than an attempt to replicate the
original hydrologic pattern. The previous dendritic or trellis network, with a
reasonably high-drainage density, was replaced by too few artificial drainage
ditches that, as well, carried too much runoff at too high a velocity. As a
result, they were armored with rock in an attempt to minimize erosion and
scour in the channels.

7.6.3   Elements of Critical Concern

The following is an outline of the most important elements of critical concern
when attempting to restore and reclaim mined landforms and to construct
stable tailing piles or embankments. The design alternatives that apply to these
elements is discussed in greater detail in the next section.

A. *Shape and Orientation of the Footprint*
   - Rectangular or near-rectangular tailing footprints do not represent
     natural hill terrain analogs.
   - The alignment of the longitudinal axis should not be parallel to the
     primary wind direction.
B. *The Slope Profile in Cross Section*
   - Linear, benched, and uniform gradient slopes are not natural ana-
     logues.
   - Transitions between tailing slopes and planar natural surfaces are not
     naturally abrupt.
C. *The Slope in Plan and Frontal View*
   - Long, flat slope faces without concavity or convexity are not typically
     natural phenomena.
   - Extensive linear alignment of slopes does not represent a natural an-
     alog.
   - Tops of tailing dumps in linear configuration without vertical profile
     relief are not natural analogs.
D. *Treatment of Plateau*
   - If that surface is not used for storage of "liquids," it should be used
     to develop natural landforms with their own hydrologic pattern rather
     than remaining a flat, artificial plain.
E. *Pits and Other Excavations*
   - Pits and other excavations left behind should provide for soft transi-
     tions to the terrain above grade or, if intended for use as a lake, have
     a designed, natural-looking shoreline.

F. *Revegetation and Reforestation*

- Linear and equidistant planting of trees does not represent a natural analog.

7.6.4 Design Alternatives

The following is a discussion of proposed conceptual alternatives to current conventional tailings disposal and storage designs. The objective is to create, from the outset, more permanent, mature, stable, and natural landscape designs that are sustainable beyond a 100-year threshold. The following are some of the design alternatives available.

7.6.4.1 The Shape of the Footprint: The typical rectangular or near-rectangular shape has no apparent redeeming qualities other than simplicity in design and surveying (see Figure 7.21). On the other hand, highly complex and irregular shapes are not necessary, may not represent natural analogs, and may needlessly complicate surveying and placement logistics.

The shape of the spoil mound should be designed so that it aids in reducing the impact of various erosional forces. The nature of these erosive forces is discussed in detail in Chapter 2. The influence of topography and selection of slope shapes that best resist erosion are discussed in Chapter 4. A shape that appears to have particular merit and a number of advantages over a rectangular or linear shape is a basic "kidney" or "double kidney" formation (see Figures 7.22–7.23). This shape most closely resembles the perimeter of a natural hillside area without creating an overly complex configuration.

This configuration has the following advantages:

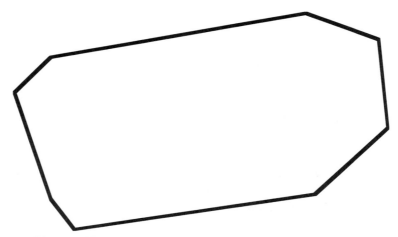

Figure 7.21 Conventional footprint with no containment area.

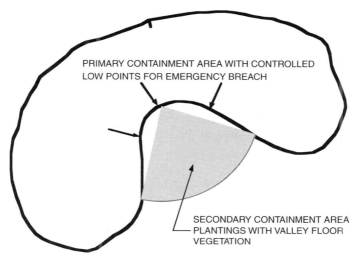

Figure 7.22 Primary containment area with controlled low point for emergency breach and secondary containment area with valley-floor plantings.

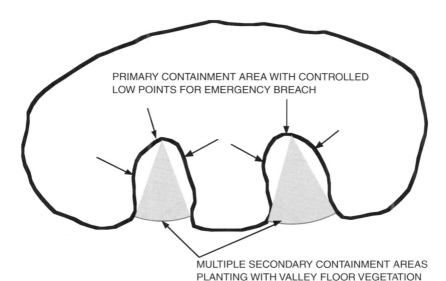

Figure 7.23 Primary containment area with controlled low points for emergency breach and multiple secondary containment areas.

1. It creates wind shadows within the structure.
2. It provides for controlled opportunities for secondary containment basins in case of a breach.
3. The shape establishes valleys and ridges, thereby creating variety in natural landforms and opportunities for establishment of diverse plant communities.
4. It blends into the flat landscape more gradually and gently, thereby creating a better transition in the landscape.
5. It represents a more natural analog for hillside terrain.

7.6.4.2 The Orientation of the Footprint: The longitudinal axis of the tailing dump should not be aligned parallel to the primary wind direction; rather, it should be shaped and positioned to create the opportunity for wind shadows. This configuration will aid in the prevention of wind erosion for at least some shape segments (see Figure 7.24). Wind-tunnel testing may be advisable to determine optimum shape and orientation.

7.6.4.3 Slope Profile in Cross Section: A slope face at a uniform angle from top to toe invites sheet flow and predictable erosion and gullying, as experience has shown. Rather than awaiting inevitable channel incision, degradation of the slope face, and formation of an alluvial fan at the toe, a slope face should be designed with a variable slope angle to forestall this evolution. In other words, the gradient of the slope face should decrease as it is descends, and it should end in an alluvial fanlike structure at the bottom.

This concave formation *anticipates* the ultimate equilibrium of the slope and, therefore, already establishes a *mature* landform from the start. The following are the advantages of this profile and shape:

1. It is geotechnically more stable (refer to Chapter 4).
2. It is less susceptible to gullying and erosion as runoff velocities are decreased even as runoff quantities tend to increase.

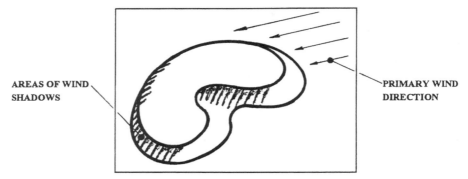

Figure 7.24 Primary wind direction and areas of wind shadow.

3. Less sedimentation can be expected because of less ongoing erosion.

4. Opportunity exists for deposition of eroded material occurs within the footprint of the spoil mound rather than allowing it to encroach onto the surrounding natural environment, e.g., streams, lakes, etc.

5. It establishes a mature landform from the start rather than awaiting the unpredictable results of an uncontrolled and random erosional and sedimentation process decades later.

6. It permits faster and better establishment of plant growth on flatter slope segments.

7. It will minimize corrective mitigation work later.

8. It blends more readily into the surrounding topography through gradual transitioning (see Figure 7.25).

The advantages of a concave-slope profile, vis-à-vis erosion and sedimentation, have been recognized previously by others (e.g., Meyer and Kramer, 1969). Various slope-profile alternatives, such as uniform, convex, concave, and complex (compound) were investigated by means of laboratory, field, and computer simulations. The results of these studies are described and analyzed in Chapter 4. The basic conclusion from these studies: Compared to all other slope forms, concave-slope profiles erode less, produce less sediment, and are more stable over time than uniform or convex slopes.

7.6.4.4   The Slope in Frontal and Plan View

7.6.4.4.1   The Rim of Plateau (Top Surface): The widespread linear configuration of the rim of current dump designs creates a stark and unnatural landform with no apparent benefit and further represents an intrusion of an artificial structure into the natural landscape. Vertical relief along the edges and perimeter of spoil mounds provide additional protection against gullying and channel incision caused by overflow from the top. The dangers of gullying as a result of abrupt changes in slope (see Figure 5.12) are discussed and analyzed in Chapter 5, Section 5.8.3 Applications of Digital Terrain Mod-

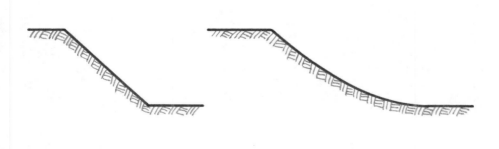

CONVENTIONAL PROFILE   PROPOSED CONCAVE PROFILE

Figure 7.25   Conventional vs. concave slope profile, showing gradual transition into surrounding natural terrain of latter profile.

eling. In addition, this variation in vertical relief along the slope rim creates a more natural and interesting visual appearance as shown in Figure 7.26.

7.6.4.4.2 Slope Form in Plan View: The use of concave- and convex-slope components is critical to landform grading slope designs. If these slope shapes or elements are employed, runoff would be dispersed in radial, instead of linear, patterns (see Figure 7.9). Water would then be naturally channeled in established and reinforced concave swales and then carried in predictable patterns across the slope face to the toe (Summerfield, 1991). This pattern would be further enhanced with diverse plant communities appropriate for concave and convex slope forms, respectively (see Figure 7.27).

Spoil mounds with level tops or plateaus are apt to develop ponded or boglike conditions. Water infiltrating into the mound from this ponded water on top can sometimes flow laterally, upon encountering a less permeable layer, and emerge (daylight) at the face. This lateral seepage can destabilize the slope and/or cause seepage erosion at the face.

A reasonable solution to this problem, other than continuously draining the top, e.g., via a system of vertical wells and horizontal drains, appears to be

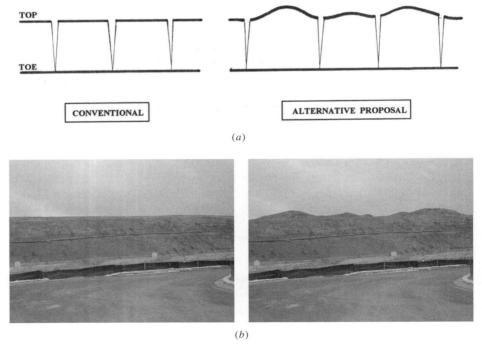

Figure 7.26 Conventional vs. alternative slope-rim designs: (*a*) schematic illustration and (*b*) photo comparison; flat linear crest versus crest topographic relief added in the form of naturally appearing ridgeline.

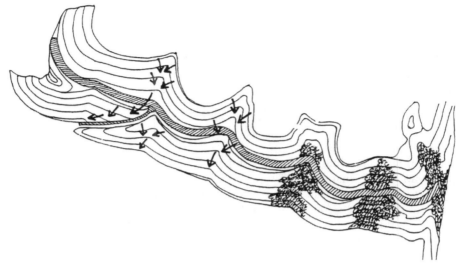

Figure 7.27 Topographic representation of landform configuration showing radial flow of water and foliage placement in swales.

a combination of topographic modifications and revegetation. High ground can be established within the shallow basin or depression on top of the mound in the form of a series of peninsulas, islands and deltas projecting into the water surface.

Phreatophytes or water-loving trees and vegetation—e.g., willows, equisetum, etc.—should to be planted around the perimeters of these projecting features. The roots of phreatophytes continuously remove water by a process of transpiration. Over time, these vegetated areas would slowly encroach further into the water area and minimize the threats to stability posed by ponding and downward seepage. Another advantage of this approach is that removal of water accumulating on top of a spoil-bank plateau before it can seep deep into the mound, also minimizes leaching of potentially toxic materials into the spoil.

7.6.4.5 Revegetation and Reforestation: The current practice of planting trees in linear rows is not desirable for the following reasons:

- It invites erosion between tree rows.
- It is not conducive to natural plant growth.
- It has an artificial and unnatural appearance.

Slopes graded with natural landforms, on the other hand, provide the necessary opportunity for revegetation, according to the principles and guidelines explained earlier in the chapter. Trees and shrubs are suitable for establishment in swales or valleys. Here they can be clustered in patterns where they thrive best and where they help with stability. On the one hand low growing groundcovers, which require less moisture, can be established on drier high ground or interfluve areas where they help prevent surficial erosion.

7.7 SUMMARY AND CONCLUSIONS

This chapter describes the essential characteristics and difference between conventional and landform graded slopes and topography. These differences show up in the shapes and profiles of the graded slopes, in the drainage systems employed, and in the manner of landscaping or revegetating the slope surface. *Landform Grading* principles can be applied to both hillside developments and to reclamation of surface mining. By applying the various alternatives, options, and combinations that are the hallmark of *Landforming,* it is possible to create a more mature and stable geomorphic landscape that future generations can accept and appreciate.

7.8 REFERENCES

Qian, X., R. M. Koerner, and D. H. Gray. 2002. *Geotechnical Aspects of Landfill Design and Construction.* Upper Saddle River, New York: Prentice Hall.

Meyer, L. D., and L. A. Kramer. 1969. Erosion equations predict land slope development. *Agricultural Engineering* 50: 522–23.

Summerfield, M. 1991. *Global Gemorphology.* Upper Saddle River, New York: Pearson.

8 Essential Design Elements for Slope Forms and Landforms

8.1 INTRODUCTION

Man-made alterations in natural landforms, whether for hillside development, highway construction, mining reclamation or landfills, in the majority of cases, will reveal that the most prominent element following alteration is the slope. Furthermore, this alteration also shows how radically different that element appears compared to its original or adjacent natural counterpart as shown in Figure 8.1.

Natural slope faces appear to possess an endless variety of forms and shapes, both concave and convex, in nature that create interest and diversity. The slope face attracts attention but, so, too, does the entire landform assemblage, of which the slope is a part, as it forms its crest line, toe boundary, as well as the entire body of the landform on which the slope borders.

A study of natural slope forms holds the key to understanding their essential characteristics and to replicating them as closely as possible during grading operations. The goal is to capture not only the aesthetic character of these slope forms but also to restore their original integrity in terms of hydrology and vegetative patterns. Desert regions lend themselves particularly well to a study of slope and landforms because little or no vegetation obscures their shapes and formational processes.

8.2 NATURAL LANDSCAPE ELEMENTS

8.2.1 Origin of Natural Slope Forms

Essentially all slope forms are products of long-term erosional and mass-wasting processes. The agents of erosion—namely, running water, gravity, and wind—begin almost immediately to reform any flat, inclined surfaces to facilitate runoff and sediment transport in the most efficient manner. Vegetation growth and distribution will accommodate themselves to the developing topography in an interactive or synergistic manner. As noted in Chapter 4, sheet runoff tends to coalesce quickly into small rivulets, and if velocity (and tractive stresses) are high enough, the runoff will incise to form large, permanent channels. These channels, furthermore, tend to descend within swales

(*a*)

(*b*)

Figure 8.1 Comparison of natural vs. man-made slopes showing radically different appearance in shape and form: (*a*) natural and (*b*) man-made.

or valleys with concave profiles. The resulting slope forms, too, will tend to have compound convex and concave surfaces as opposed to simple planar shapes.

8.2.2 Natural Drainage Forms

Natural slopes typically exhibit a variety of concave and convex forms and as a result run off is controlled and channeled across the slope face in a curvilinear manner that follows natural drop lines in the slope. Channel curvature increases length and decreases gradient and velocity. Artificially created slopes, on the other hand, do not replicate this pattern but, instead, create uniformly inclined planes that promote erosive sheet flows and rilling as shown in Figure 8.2. To interrupt these erosive flows down these slopes, the runoff has traditionally been captured on horizontal terraces or benches and then diverted directly downslope in hardened channels or pipes. This approach has resulted in the practice of regularly spaced horizontal terraces, terrace drains, vertical down-drains, and interceptor drains.

8.2.3 Natural Vegetation Patterns

Every slope and landform exhibit not only topographic features and runoff patterns but also distinct plant communities and areal distributions. This veg-

Figure 8.2 Planar slope face with sheet flow and development of rills. Cut slope along main access and haul road, Redwood National Park, California.

etation pattern is not random but, instead, is based on the availability and retention of moisture in a given location, as determined by the slope's topography and runoff patterns. This dependent relationship is particularly evident in regions with seasonal rainfall regimes. Trees and major shrubs are largely concentrated in concave swales. This is where the largest concentration and retention of moisture occurs and where woody vegetation is generally well adapted, as shown dramatically in Figure 8.3. Grasses and herbaceous ground cover, on the other hand, are typically found on the convex, interfluvial slope areas, where less moisture is available to sustain them. Typical distributions of plant communities or vegetation patterns on natural slopes are illustrated in Figure 8.4.

8.3   BASIC SLOPE FORMS—"THE ARCHITECTURE OF SLOPES"

8.3.1   General Observations

Careful observation and study of a variety of grading plans show that slope configuration can significantly impact the building pad configurations above and below the slope, affect road alignments, and ultimately influence the placement of all structures. This conclusion is based on a four-year-long study, observations, and a photographic survey of slope and landforms from different regions of the world, as well as examination and analysis of topographic maps of natural landscapes and slope forms. What, at first glance, appeared to be a never ending variety of slope forms covered with an infinite variety of vegetation proved surprisingly amenable to categorization and to quantifiable description.

The results of this study allowed various slope forms to be synthesized into a series of reasonably well defined and repetitive shapes. The main variable was their size and scale in relationship to the overall landform. The finding of greatest significance for future applications in man-made earthwork or grading plans was that these slope elements or forms were not dependent on soils nor on climate. Instead, they proved to be universal in their nature and occurrence. Their universal character lend themselves to replication for any potential grading application.

The following is a list of those repetitive forms that not only give natural slopes a pleasing and harmonious appearance but, at the same time, function to control runoff effectively and to provide an ideal environment for plant establishment and growth.

8.3.2   Ridges and Swales—Perpendicular to the Slope Crest

Ridges and swales are one of the most common and pronounced elements of slope form. Each concave and convex form is located perpendicular to the

Figure 8.3 Tree establishment in channel cut in rocky slope, where some moisture is available. Note absence of any vegetation in surrounding rock slope.

slope crest and generally spaced fairly equally across the slope. This spacing or distribution limits the tributary drainage area in each swale and controls the amount of slope runoff. This action, in turn, minimizes erosion, while simultaneously stimulating plant growth and diversity across the entire slope. Examples of this slope form, which occurs in diverse geographic regions of the world, are shown in Figure 8.5.

8.3.3 Ridges and Swales—Diagonally across the Slope Face

Ridges and swales located diagonally across the slope face are the simplest form of natural runoff interception and among the easiest to replicate in a grading plan. The diagonal placement provides a longer path for runoff to flow downslope and thereby limits channel gradient and resulting flow velocity. An example of this slope form is shown in Figure 8.6.

(a)

(b)

Figure 8.4 Distribution of woody vegetation (trees and shrubs) and grasses in a natural hillside. Woody plants tend to cluster in swales while grasses grow in drier interfluves: (a) California, United States, and (b) Alberta, Canada.

(a)

(b)

(c)

Figure 8.5 Examples of ridge and swale elements perpendicular to the slope face in diverse geographic regions of the world: (*a*) Santa Ana Canyon, California; (*b*) Sabana Mountains, Brazil; (*c*) Canadian Rockies, Alberta, Canada.

Figure 8.6 Example of ridge and swale elements located diagonally across the slope face Swarthout Valley, California.

8.3.4 Ridges and Swales—Curvilinear across the Slope Face

Ridges and swales located in curvilinear fashion across the slope face function in much the same manner as diagonal placement but with an additional aesthetic or visual benefit. These shapes function in a graceful and natural way to manage runoff across the slope face by collecting and carrying water to the bottom of the slope without the need for regimented, horizontal terracing and vertical down-drains typical of man-made slopes. Furthermore, channel sinuosity lengthens the channel in the swale and reduces gradient even more compared to diagonal placement. Examples of this slope form are shown in Figure 8.7.

8.3.5 Elbow Shapes across the Slope Face

Nature demonstrates in another unique way how it controls runoff, by putting "elbows" across the slope face. An elbow is a ridge and swale combination with a kink or bend. At the upper region of the slope, where runoff quantities and velocities are small, the swale is still perpendicular to the slope face. However, at about the slope midpoint, the flow in the swale is slowed by about a 45-degree angle turn or elbow bend. This results in a more diagonal descent and lower gradient, with correspondingly less erosive flow to the toe of the slope. Examples of this slope form are shown in Figure 8.8.

8.3.6 Pyramid- and Cone-Shaped Slope-Face Elements

Pyramid- and cone-shaped slope-face elements are also common slope elements that are easy to replicate in grading designs. The main difference between these two elements is that the cone shape has a rounded face and the pyramid a flat one. One can also observe in nature a wide variety of these shapes, for example, wide base, narrow base, partially up the slope face, or totally covering the primary slope face. Figure 8.9 illustrates the variety of pyramids and cones as they appear in nature and shows their forms and proportions.

8.3.7 Wishbone Configurations

A wishbone-shaped slope element is a dramatic yet fairly common form that consists essentially of two descending ridgelines separated by a central swale. The examples in Figure 8.10 illustrate the wide range of sizes in which this form appears, ranging from a 30-foot high shape in Death Valley to the 700-foot tall "structure" in Mt. McKinley (Denali) National Park, Alaska. Other intermediate sizes can also be observed. It is interesting to note that, although differing in size, each example is similar in character and proportion in spite of the fact that each was formed under completely different climatic conditions and with very different soils and bedrock.

(a)

(b)

Figure 8.7 Examples of ridge and swale elements located in a curvilinear fashion across the slope face: (a) Weir Canyon, California, and (b) Maui, Hawaii.

(a)   (b)

(c)

Figure 8.8   Examples of an elbow-shaped slope element located across natural slope faces: (*a*) Yellowstone, Wyoming; (*b*) Swarthout Valley, California; and (*c*) Santa Ana Mountain, California.

8.3.8   Convex Ridges and Concave-Foot Slopes

Few, if any, slopes in nature have sharp, angular crests or toes. The crests of mesas or buttes that are capped by hard rock layers are an exception to this rule, but foot slopes are generally concave in shape (see Figures 4.15 and 4.16).

As discussed in Chapter 1, numerous geomorphic studies of natural landscapes have shown that many soil-mantled hill slopes have compound, curvilinear shapes. Studies by Hancock et al., 2003, for example, have shown that soil-mantled, fluvial erosion–dominated catchments generally have convex upper hillslope profiles with concave profiles developing further downslope (see Figure 4.13).

8.3.9   Compound and Composite Shapes

Compound and composite shapes are either relatively unique creations or are combinations of some of the preceding elements. Examples of some possible combinations are illustrated in Figure 8.11. Figures 8.11*a* and *b* shows a remarkably complex yet common reoccurring pattern, which could be inter-

(*a*)

(*b*)

Figure 8.9   Example of cone- and pyramid-shaped slope elements located on a natural slope face: (*a*) Santa Ana Canyon, California; (*b*) Swarthout Valley, California.

(a)   (b)

(c)

Figure 8.10 Examples of wishbone-shaped slope elements located in diverse geographic regions: e.g., (a) Death Valley, California; (b) Mt. McKinley (Denali) National Park, Alaska; (c) Santa Ana Mountains, California.

preted as two elbow-shaped or curvilinear swales, at either side, wrapping around a slope surface that itself exhibits other slope elements, such as alternating perpendicular ridge swales. Figure 8.11c shows the central part of a slope flanked by two cone-shaped elements—one developing into a wishbone configuration. This configuration is essentially the same as shown previously in Figure 8.11b, but it occurs in an alpine setting with higher rainfall. Figure 8.11d shows a ridge with a series of secondary ridge swales and ridges oriented perpendicularly to the main ridge. A large swale and ridge cuts diagonally across the slope as well. This results in truncation of the secondary ridges and the formation of pyramidal- or cone-shaped slope forms as well.

In considering all these slope shapes and forms for replicative design, it must be kept in mind that they can also vary in certain modifying details beyond their basic outlines or themes. These modifying details or characteristics provide further opportunity for variety and creativity in grading designs. Modifications of the basic elements described previously consist of the following:

8.3.9.1 *Degree of Roundness or Angularity:* Most slope forms have a further variation in their exterior structure. Protruding, convex ridgelines, for example, either perpendicular to the main ridge crest, curvilinear, or diagonal

(a)

(b)

(c)

(d)

Figure 8.11 Example of composite or compound slope elements: (*a*) Anaheim Hills, California; (*b*) Banning, California; (*c*) San Gabriel Mountains, California; (*d*) Death Valley, California.

across the face can display roundness while others can be quite angular or jagged.

8.3.9.2   Width: Width refers to the size of the base of a particular slope form. Different slope elements can differ in size but still exhibit the same basic shape or form.

8.3.9.3   Height: Refers to height across the slope face, either partial coverage or coverage that extends to the top or beyond the crest of the main slope. The latter, in effect, helps to create relief and results in an irregular ridge line.

8.3.9.4   Proportion: This factor is probably the most significant one from the standpoint of design, because it relates the height of a given slope form to the amount of indentation, or concavity, from a tangent line drawn across the toe of slope. Proportion will have a direct impact as to how visible the form will be to an observer from a given distance. This issue will be discussed in more detail in Chapter 9.

8.4 THE ROCK ELEMENT AS PART OF THE NATURAL LANDSCAPE

Rock ledges, outcrops, and boulders are also part of the natural landscape. In arid or desert landscapes, these rock elements may constitute the central feature of the landscape. These rocky elements lend visual interest and diversity to a landscape. In the landform grading approach, many of these rock features are preserved intact where possible. For example, instead of burying all boulders in fills or blasting, braking, and crushing them, many boulders are instead selected, hauled, and pushed on to holding areas for later incorporation into the graded landscape.

The choice of locations and frequency of placement of boulders and rocks in a grading or landscape design requires some sensitivity so it will mimic as much as possible the natural landscape. Placement of rocks and boulders (in a natural manner) in graded valleys and swales can contribute further, aiding in erosion control. These kind of placements, or replacements, of rocks and boulders not only enhance the aesthetics but go a long way toward restoring the integrity and character of the original landscape. Additional information about the placement and handling of rock elements is provided in Chapter 9, which deals with implementation of landform grading plans.

8.5 REFERENCE

Hancock, G. R., R. Loch, and G. R. Willgoose. 2003. The design of post-mining landscapes using geomorphic guidelines. *Earth Surface Processes and Landforms* 28:1097–1110.

9 Implementation of the Landform Grading Plan

9.1 REQUIREMENTS FOR SUCCESSFUL IMPLEMENTATION

A landform grading plan is subject to most of the same requirements as a conventional plan; namely, it must provide for drainage and conveyance of surface runoff and take into account possible geologic hazards and potential subgrade problems. These requirements were described and discussed in Chapter 6. A landform grading plan must also address additional requirements and considerations. These include initial site planning, meetings with regulatory agencies, geotechnical analyses, the grading phase (including retraining of grading personnel), landscaping or revegetation protocols, and cost analyses.

9.2 OBSTACLES TO IMPLEMENTATION

There are five major hurdles to implementation of a landform grading and revegetation plan. These hurdles include the following:

1. The frequent rigidity in the ordinances and standards of approving agencies that prevent granting deviations and waivers from expected norms.
2. The tendency of civil and geotechnical engineers, professionals with primary design responsibility, to design strictly along traditional and well-established lines.
3. The developer's fear of potential cost increases or a mining company's concern about the bottom line if this new approach is adopted.
4. The mass-grading contractor's concerns about limitations on workers and machines that this new approach might incur.
5. The developer's concern about a possible reduction in the yield of usable pad areas or individual home sites in the case of an urban hillside project. A pad is understood to mean a graded, level building site ready to accommodate conventional foundations for residential, commercial, educational, municipal or other buildings.

9.3   IMPLEMENTATION STRATEGIES

9.3.1   Land Planning and Initial Site Design

Land planning and initial site design is the stage where a change from the traditional approach of Step 1, planner laying out the land plan independently first, and Step 2, engineer, reworking the land planners design to meet engineering criteria needs to be made. Cooperation and coordination between these two professions is absolutely essential from the start as opposed to the current practice of land planner first, and engineer second, reviewing the planner's efforts. Best results are achieved when both work literarily side by side, in the same room. Only then can results be achieved at the outset that are both aesthetically pleasing and environmentally responsive as well technically functional and economically feasible.

The key to successful incorporation of *Landform Grading* concepts into any hillside development plan is to make provisions for irregular, variable slope patterns (rather than linear, single-slope ratios) and building pads that conform to areas with landform-shaped perimeters in the initial stages of the conceptual design (see Chapters 7 and 8). This initial planning will minimize and help allay concerns about the potential loss of usable building sites or pad areas. This approach should be carried forward all the way from a conceptual preparation phase, through initial design stages, to construction in the field.

The objective is to incorporate naturally appearing hillside slopes and landforms into the grading plan such that there is minimal or no loss of overall useable area and where any loss is offset by other benefits. Concave, flatter-slope portions, for example, are laid back into spaces between structures (i.e., at $2\frac{1}{2}:1$ and $3\frac{1}{2}:1$ or flatter). These areas simultaneously become part of the total landscaped requirement for the entire site. Convex portions, on the other hand, are steepened (i.e., at $1\frac{1}{2}:1$ and $2:1$) in front of building locations to create more level pad space. Examples of these site planning and grading concepts are illustrated in Figure 9.1.

In all these efforts to implement landform grading, the most difficult transitions to make from the designer to the grading contractor to the landscaper was the change not only from the linear or even the curvilinear (via contour grading) but to the creation of actual landforms that entail abrupt changes from concave to convex. There is always the temptation to fall back on the "ocean-wave effect," imaginary rolling terrain rather than true landforms.

9.3.2   Meetings with Regulatory Agency

The first order of business in implementing a landform grading and revegetation plan is to seek a meeting with the appropriate department heads of the relevant regulatory agency. The main purpose of this meeting is to obtain basic approval for implementation, so there will be no surprises when plans

Figure 9.1 Illustration of landform graded project showing placement of buildings and structures that are integrated with the slope forms.

are submitted that deviate significantly from expected norms. What to avoid, in this regard, is having an untrained city inspector arrive at a freshly graded slope and find an elbow-shape, which is a complete violation of the city's regular standards. Resolution may take a series of excited phone calls until the city engineer is able to clarify the situation and explain that all was going well and according to plan.

In-house planner, civil engineer, engineering-plan checker, and field inspector, all are primary positions requiring a complete understanding of the concept.

Sometimes it may be necessary to institute a pilot study for a smaller, more limited program to convince the agency of its effectiveness and safety. This typically would extend, at least, through one rainy season because of concerns about potential erosion.

9.3.3 Allaying Engineering Concerns

Initial discussions should be held with civil and geotechnical engineers assigned to a project about slope shapes, particularly about the use of concave-slope profiles. The use of this slope shape or other irregular slope profiles is likely to be met with some skepticism at first for a variety of reasons. Most civil or geotechnical engineers are accustomed to dealing with planar slopes with unvarying gradients. It may be necessary, therefore, to begin by pointing out the superior performance of concave profiles compared to linear profiles with regard to both superficial and mass stability (see Chapter 4).

The concern about concave slopes is more warranted in the case of three-dimensional or bowl-shaped hillside swales. This geometry can result in elevated groundwater tables or pressures at the lower end of the bowl, which in turn can decrease mass stability and/or initiate channel erosion as explained in Section 2.11—Slope-Stability and Channel-Erosion Thresholds. This problem can be avoided, however, by breaking the hillside up into a series of smaller, intermediate swales so that the tributary drainage area to each swale area is limited. In addition, the swales are planted with deeper rooted, woody vegetation that can handle additional water and that also improves mass stability (see Chapter 3). It is also important to note that landformed slopes are typically only 30 to 100 feet in height and rarely, if ever, higher than 150 feet. This height limit also limits the *topographic ratio* a/bM which in turn limits flow convergence and the tendency of the slope to saturate (see Eq. 2-12, Section 2.11.3—Slope-Stability Thresholds).

9.3.4   Geotechnical Engineering

Greater attention must be paid to geotechnical requirements for two reasons: (1) engineering geologists and geotechnical engineers are not accustomed to routinely analyzing the stability of nonplanar-slope forms and (2) landform grading plans result in more frequent variation in slope orientation and slope-face inclination. The influence of nonplanar, compound slope shapes on both surficial and mass stability was examined in detail in Chapter 4. With the exceptions noted previously, namely, the tendency for elevated groundwater tables or pore water pressures at the lower end of bowl-shaped drainages, concave-slope profiles generally have a beneficial influence on minimizing erosion and improving mass stability. The geomorphic evolution of slopes was examined in Chapter 5; and equilibrium slopes tend to have compound (convex and concave) as opposed to planar shapes.

In addition to the aforementioned advantages of nonplanar-slope forms, greater freedom with orientation and inclination of slope faces can improve stability in cases of adverse bedding conditions. On the one hand, continuously variable horizontal- and vertical-slope orientations in relationship to the natural bedrock conditions, under cut conditions, may require more stability analyses to be undertaken. On the other hand, the continuously variable angles and slope orientations may also avoid the need for buttressing and stabilization if they result in favorable orientations with respect to bedrock attitudes.

9.3.5   Introduction of Concept to Grading Designers

Serious implementation of the *Landform Grading* concept begins with its introduction to the grading designers. These are the professionals who will translate the concept into plans from which construction will proceed. The bulldozer operator in the field, ultimately, executes the plan the grading designer has drafted in the office. If the grading designer's training, background, and experience are based largely on linear and planar formulas, then reorientation of some significance will be required. This introduction and reorientation can proceed in a variety of ways, including: (1) a slide show depicting the evolvement of the concept from examples in natural landscapes; (2) a demonstration of how to best mimic these examples using the landforms described in Chapter 8; and (3) a review of past landforming projects using detailed photographs, topographic maps, concept drawings, and, as well—when all else fails—a three-dimensional model, which almost always has the most impact. Often a simple sketch or clay model will help people visualize different types or shapes of slope elements, drainage ways, and their placement on a hillside, as shown in Figure 9.2.

9.4 PLANNING AND SURVEYING REQUIREMENTS

9.4.1 Planning Requirements

Large-scale development plans (of 500 acres, 1000 acres, or larger) often begin life at scales of $1''=300'$ or $1''=200'$ and smaller sized development plans at $1''=100'$ or $1''=60'$. Final grading and construction plans are typically drawn to a scale of $1''=40'$; for smaller projects, or where more detail is required, plans are drawn at $1''=30'$ or $1''=20'$ scale. The basis for such plans is always the topographic map that depicts the underlying contours and elevation lines at various elevation intervals, depending on the steepness of the topography. An example of a topographic map and grading plan for a landformed slope is shown in Figure 9.3. For 100- or 200-scale plans, the contour interval can be as much $10'$ or more for 40 scales; it can range from $5'$ to $1'$, again depending on the steepness of the terrain. As a general principle, it should be kept in mind that the closer together the contour lines are spaced, the steeper the terrain—the further apart, the more level.

During the design phase, it is sometimes helpful to construct three-dimensional models that replicate the natural slope and landforms to facilitate visualization and to transfer and replicate these forms onto plans. After several such training sessions, most talented designers adapt fairly quickly to the *Landform Grading* concept and should be able to execute such designs more or less routinely. Models are also particularly effective with lay persons, both the public in general or reviewing and approving political bodies.

9.4.2 Surveying Requirements

It should be kept firmly in mind that that there is a fundamental difference between surveying for landform graded slopes versus for conventional slopes, because the objectives of each differ considerably. In the conventional ap-

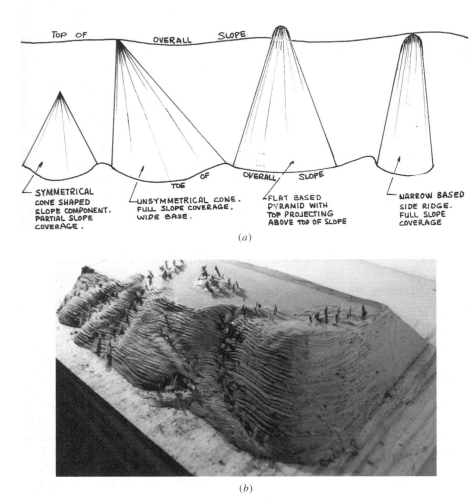

(a)

(b)

Figure 9.2 Landform grading design aids: (a) simplified reductions of some natural slope elements, and (b) clay model to help visualize landform grading design.

proach, the goal is to provide strict linear control in the horizontal direction and to assure that in the vertical plain a uniform design-slope ratio (2:1, 1.5:1) is strictly maintained. Tops and toes of slopes are precisely controlled as straight lines and with no vertical relief.

In the case of landform graded slopes, the objectives are not toward linear, mathematical precision but, rather, the replication of natural forms, which are characterized by irregularity (see Figure 9.3). Ideally, the more irregular and nonplanar the slopes, the more acceptable they are likely to become; during inspections, this design goal should be duly recognized.

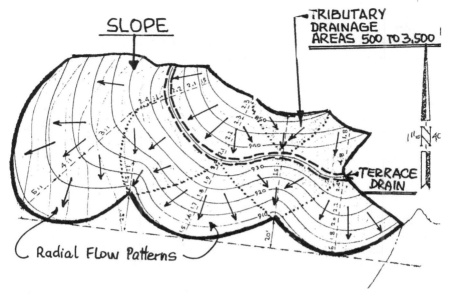

Figure 9.3 Topographic map and grading plan for a *landformed* slope.

In the early stages of implementing landform grading, too much survey control was often provided to the extent that a slope ended up with a "forest of stakes." Survey personnel soon realized that far less staking was needed for essential controls at the top and toe of a slope. The basic alignments for the concave and convex lines that were needed to define intervening topographic forms on the future slope face were interpreted by a talented and trained grade checker.

With global positioning system (GPS) capabilities, survey control has become even more simplified as well as less expensive. A sample of a conventional survey approach for a landform graded slope is provided in Appendix 1.

9.5   THE GRADING PHASE

9.5.1   Retraining of Grading Personnel

The initial approach to retraining and preparation of the grading contractor for implementation of a new concept was to discuss its characteristics and requirements with the principals and management of the firm. The assumption was that they would pass on this information to their field staff, superintendents, grade checkers, and equipment operators. For reasons that are not entirely clear, messages often failed to get through. A common misconception that apparently stymied communication was that sharp transitions or grade changes in slope forms, from convex to concave, could not be executed by bulldozers and scrapers.

This setback prompted a new approach to training of grading personnel. The new approach is perhaps best explained through the following anecdote:

> Sample photos and a small clay model of the desired slope forms were taken out to the grading project. The dozer operator was called down from his machine; the grade checker joined in for an impromptu meeting on-site at the bottom of a slope slated to be transformed into a slope with the requisite shape and form. After seeing and understanding exactly what was expected of them, the response of the dozer operator and grade checker was a prompt and assured: "Sure we can do that, why didn't somebody tell us."

> Subsequently, it was possible to see at many project sites a dozer operator stopping after finishing the earth grading, getting off the machine, glancing along "his" slope and nodding his head with satisfaction! Upon completion of a major mile-long, 75-foot high landformed and graded slope, some of the grading contractor's staff looked back along the slope profile as it was highlighted by the setting sun and the forming shadows and commented, "That is really awesome!"

So, instead of being an obstacle to landform grading, the dozer operators became enthusiastic and willing partners.

9.5.2   Ground Preparation

Preparing the ground for cut and fill operations is typically performed in the traditional manner. All remedial work, such as removal of deleterious material (see Chapter 6), is carried out. A system of subdrains is installed to intercept and collect seepage and spring water. Landslide debris is either removed or stabilized in place through the use of various techniques such as total removal, unloading and buttressing, shear keys, tie backs or piles (see Section 6.4.2.3— Unstable slopes and landslides)

9.5.3   Slope Construction

Grading techniques first begin to diverge significantly from conventional practice during this phase of the work. Stabilization fills and/or buttresses are constructed in cases where cut slopes expose adverse bedding conditions. This work is supplemented by the construction of subdrain galleries at the so-called "false cut," or final back-cut slope face in order to control subsurface seepage, to conduct this flow safely away, and to prevent saturation of buttress or stabilization fills.

9.6   FILL CONSTRUCTION AND COMPACTION CONTROL

Fill construction and compaction control for landform grading designs are basically the same as conventional grading plans. The main concern in both cases is to achieve adequate compaction (density) near the slope face where there is less lateral restraint. Compaction testing in the field may occasionally have to be done at more frequent intervals. Additional costs are difficult to determine, as they are highly variable but based on past experience, costs can be at most between 1 and 2 percent higher than the usual geotechnical services associated with compaction control.

9.7   CONSTRUCTION OF VALLEY OR "DAYLIGHT" FILLS

The placement and construction of valley fills are required when grading for hillside development, among other reasons, reaches the topmost portions of topographic uplands and canyons and swales become narrower. These locations will not yield more developable land, but they often need to be filled upstream for considerable distances just to allow upstream drainage to "daylight," that is, to provide for downstream gravity-outlet flow.

These fills have traditionally been constructed as level fills with steep downstream planar faces that encroach into the natural landscape as an artificial landform that over time will remain a "scar," because it is visually incongruent with the surrounding terrain (see Fig. 9.4). Instead, in the landform grading approach, these areas are reshaped as valley fills by placing benched fill into the side canyon slopes while at the same creating a meandering or sinuous flow line in the newly created valley floor atop the benched fill to establish a more natural landform. Alternatvely, the fill can be placed in the canyon to replicate the original topography and drainage pattern but at a slightly higher elevation, as shown schematically in Fig. 9.5.

Once such areas have reestablished their vegetative cover, they will no longer be distinguishable from the previous or surrounding natural terrain. The meandering valley floor will at the same time encourage excellent habitat opportunities for diverse shrub and tree growth by providing concentrated flow and greater moisture availability.

9.8   SLOPE-DRAINAGE DEVICES

Sometimes code or ordinance requirements for the conventional installation of benches and hardened drainage devices cannot be avoided, even when applying a landforming approach. Under these circumstances, the following techniques can be used to give these devices a more natural appearance and to enhance the overall aesthetics of the slope.

As is readily recognized, the design form of the slope face is not only critical to its appearance but so too is the placement and design of drainage-control devices. These devices generally consist of terrace drains, down-drains, interceptor drains, or other drainage devices that are constructed from

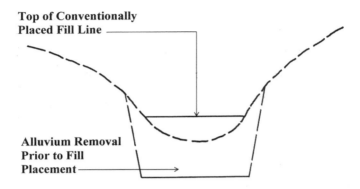

CONVENTIONAL DAYLIGHT
FILL CONSTRUCTION

Figure 9.4 Schematic illustration of conventional approach to placement of fill in valleys or canyons.

concrete or gunite. These drainage facilities have historically been considered part of an effort to provide stability to the slope face and prevent erosional damage from runoff.

9.8.1 Terrace Drains

Drains are typically less visible when viewing a slope face from the bottom where only the graded terrace or bench can be observed. In the traditional approach, the drainage channels are placed on the rigidly constructed terraces that run linearly along the length of the slope and slightly off contour with a slight inclination of 5 percent to collect and discharge runoff from the slope face above into a down-drain device. In the landforming approach, terraces and concrete terrace drains can often be eliminated as linear sheet-flow drainage patterns are replaced with radial flows (see Figure 7.6). This change should always be reviewed and approved by the geotechnical engineer with regard to soil erodibility and overall stability.

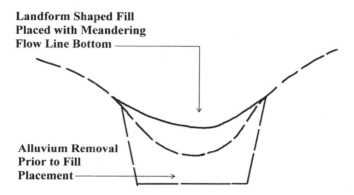

**LANDFORM/VALLEY SHAPED
DAYLIGHT FILL CONSTRUCTION**

Figure 9.5   Schematic illustration of landform grading approach to placement of fills in valleys or canyons.

9.8.2   Down-Drains

A typical down-drain is constructed from gunite and resembles a flume. This drainage device is prominent and highly visible from any direction and, as such, is unsightly on any slope face designed in the traditional form. And it remains exposed and unsightly even after the landscape is in place (see Figure 7.3). It conveys the water from the terrace drains to the bottom or toe of the slope. In spite of their negative visual impact, down-drains have historically been engineered so that they are placed on the most prominent and exposed positions of the slope, namely, at the center of the overall slope face.

Landform grading counteracts this visual impairment in two ways. The simple solution is to convey the flow, diagonally across the slope face, in a channel formed by a natural linear ridge and swale (refer to the natural-slope analog in Figure 8.6). Alternately, and even more "naturally," these flows can be collected in channels that descend a slope face in a curvilinear manner (see Figure 8.7), or they are tucked away out of site in curvilinear and elbow-shaped swale and ridge elements that cross a slope face (see Figure 8.8).

Actual examples of these much more complex and sophisticated drainage designs in which the slope face has been shaped into these natural analog or slope elements, such as a curvilinear ridge and swale or elbow-ridge and swale combination, are shown in Figures 9.6 and 9.7.

Notice how these drains fit into the landformed slope face in each case. The drains are effectively hidden from view, and in high visibility areas, they are additionally treated with natural river rock, giving them the appearance of a streambed. This treatment places these devices in harmony not only with the overall slope design and the plant material but as such they become aesthetically and environmentally responsive elements.

Care must be taken that inexperienced designer do not follow a common temptation and place these down-drains in the center of major concave swale elements. This would defeat the purpose of supporting the dense and deeply rooted plant growth planned for the same areas as runoff would be intercepted and carried off without ever reaching the vegetation.

9.8.3   Interceptor Drains

Interceptor drains are so-called perimeter drains that surround the manufactured slope (typically a cut slope) where it joins the natural slopes. Interceptor drains prevent runoff from the adjoining natural areas to encroach upon man-made slopes. When used to intercept flow or runoff from a natural slope to a cut slope, they are also at times referred to as brow ditches. Interceptor drains can often be eliminated in the landforming approach, because significant graded transition zones between the man-made slope face and the natural

Figure 9.6   Example of land development project drainage design using a curvilinear-ridge and swale combination.

Figure 9.7 Example of land development project drainage design using an elbow-ridge and swale combination.

slope are purposely created. These transitions eliminate the abrupt increase (or decrease) in runoff velocity that are the principal causes of erosion and sedimentation. Where they must remain, they too can be incorporated into the natural landform elements on the perimeter and the transition area of a slope and as well be treated in the same manner as the down-drains.

9.8.4   Toe Drains

Toe drains, normally constructed from concrete, are devices that run along the bottom of the slope. They are designed to intercept runoff from the lowest part of the slope to prevent ponding at the toe and to divert runoff away from the slope and structures and other improvements on building pads located at the base of slopes. The flow is typically diverted to the street, an underground storm-drain pipe or an open channel.

9.8.5   Hardened Drain Limitations

The major drawback of the application of hardened slope-drainage devices is that they derange the natural slope hydrology and withdraw water from the slope face that could otherwise support plant material. This runoff can sustain natural vegetation on a long-term basis and, as such, can (and has) provided on-site mitigation in the case of the "taking" or loss of natural areas during grading operations. Therefore, this benefit is provided within the project boundaries without the need to establish an off-site mitigation area. The case study in Chapter 12 of the Hollywood Hills mass-grading project describes a successful application.

9.9   REVEGETATION

9.9.1   Conventional Landscaping vs. Revegetation

Once a hillside has been mass graded, the next step (and the normal procedure) is to plant and seed the bare ground with trees, shrubs, and ground cover. One of the main purposes of this landscaping treatment on bare hillslopes is to protect the ground surface against rainfall erosion and runoff, to reestablish root structures, and to hide visual scars. Landscaping is normally applied in conjunction with surface-drainage devices for erosion control.

Historically, this landscaping treatment evolved in parallel with traditional grading methods. The same rigid thinking about appropriate grading designs also carried over into spatial distribution of plant materials. Little concern was paid to planting designs that were compatible with slope hydrogeology. Instead, the tendency was to specify uniform coverage—i.e., a rigid, evenly spaced distribution on a slope face to achieve total plant coverage. This entailed promulgating specifications simply in terms of quantity of seed or number of stems per unit area. Equal spacing of trees and shrubs was, and still is, the standard as demonstrated in Figure 9.8.

This rigid, uniform landscaping approach can further intensify an already monotonous conventional grading design with its planar slopes and unvarying gradient as illustrated in Figure 9.9. This figure also illustrates the civil engineer's tradition of placing the concrete downdrain structure in the most visually exposed position possible. Apart from its aesthetic flaws, this conventional approach is also less effective than a landform-revegetation treatment in which plant material is carefully grouped and spaced according to what type of plants are best suited for different locations on a slope.

9.9.2   Landform Revegetation

The selection and spacing of plant materials in landform-revegetation designs must be compatible with hillside topography, geology, and hydrology. Lessons can be learned, in this regard, by examining and photographing plant associations on natural slopes. Plant materials requiring more moisture—namely, trees and shrubs—tend to be clustered into areas where moisture concentrates and where evaporation is slowed, that is, in swales and valleys. Grasses, on the other hand, tend to be concentrated on the drier convex slopes or interfluves. This occurrence is particularly apparent in the semiarid regions of the

Figure 9.8 Conventional landscaping installation showing uniform distribution of trees and shrubs on a slope in a grid pattern.

Figure 9.9 "Plants on parade," an example of a conventional landscaping treatment employing rigid spacing and/or uniform coverage and rectilinear placement of down-drain in a highly exposed position.

Western states, where rainfall is seasonal (10 to 20 inches or less annually). The distribution of native trees and grasses in this manner create patterns that are stable, in harmony with the surrounding environment, and pleasing to the eye.

Landform grading designs (with their convex and concave slope shapes) constitute an inviting topographic framework for such natural vegetation patterns. There is an additional hydrologic bonus, moreover, to these compound slope shapes. Landform grading also reconfigures slope-drainage patterns from one of sheet flow to one of radial flow (see Figure 7.6). This permits the diversion and concentration of both natural rainfall and irrigation to specific areas.

Accordingly, as part of the "revegetation" process trees are grouped within concave areas and, where environmentally appropriate, clustered in areas receiving the most moisture. Shrubs are heavily concentrated along flow-drainage lines of each swale and thinned to either side. The final result is a man-made landscape that looks natural and where plant material locations and distribution are in harmony with the hillside hydrogeology, as shown in Figure 9.10.

There are further benefits to revegetation in combination with landform grading, namely, irrigation needs are reduced. Reduced irrigation occurs because:

- Concentration of runoff in concave swales, as a result of radial drainage patterns, provides the most moisture to plant types, namely, trees, that need the most moisture.

Figure 9.10   Example of a landform-revegetation application.

- Better absorption by plant roots occurs with slower water velocity, resulting from flatter slope inclinations in swales near the lower half of the
slope. Reduced velocity and/or depth flow also means lower tractive
stresses and erosion losses (see Chapter 4, Section 4.3.3, Surficial Erosion).

- Reduced evaporation results from prolonged periods of shade for plants
growing on adjoining convex areas.

The dramatic the effect that landform shaping a slope can have on the soil-
moisture regime and revegetation is depicted in the photos in Figure 9.11.
The convex portions dry faster while the concave swales remain saturated.
Landform graded slopes preserve moisture in the swale areas and speed germination. This grading plan prevents erosion from occurring in concave
swales, even though they receive some concentrated flow, because the tributary drainage in each swale is kept to a minimum.

9.10   THE APPLICATION OF WATER ON THE SLOPE FACE THROUGH IRRIGATION

9.10.1   High-Pressure Spray Method

Traditionally, water to sustain the plant material is applied though a pressur-
ized spray irrigation system consisting of primary and secondary pipelines,
feeding risers and sprinkler heads, all placed on the slope face as shown in
Figure 9.12.

This is probably the most wasteful way to dispense water; because during
the spray period, large amounts are wasted as runoff or is wind driven to the
streets and other unintended areas. Another large percentage disappears in the
evaporation process.

9.10.2   Low-Pressure Spray Method

There are other means available to dispense water, such as the low-pressure
spray or drip system (shown in Figure 9.13). Originally developed for agricultural purposes on flat or gently sloping ground, particularly in arid regions
where water is at a premium, it can be adapted to slope irrigation. Unfortunately, this system has in the past been vulnerable to rodents, who upon
discovering the source of water bite into the lines causing continuous leakage.
This method will continue to be plagued by these problems until manufac-
turers are willing to change their product and provide for harder pipe lines.

9.11   PLACEMENT OF ROCKS AND BOULDERS

More recent developments have been encroaching into even more difficult
terrain, not only exhibiting the steepness of the terrain but also highlighting

(*a*)

(*b*)

Figure 9.11 Landform-graded slope with a uniform hydromulch application: (*a*) After 3 inches of intense rainfall. (*b*) View showing quick germination and absence of erosion in swale bottom.

Figure 9.12   Example of high-pressure spray irrigation.

Figure 9.13   Example of low-pressure spray irrigation.

the presence of individual or groupings of surface boulders, rock capping, rock outcroppings, and rock ledges. Once grading has been completed, excavations are made on fill slopes, in graded valleys, and other suitable portions of the graded terrain. And boulders are either pushed by dozers or set by crane or backhoe into the excavation, providing for partial burial. An example of the placement of rocks and boulders into a graded slope in a hillside housing development is illustrated in Figure 9.14.

(*a*)

(*b*)

Figure 9.14   Placement of rocks and boulders into a graded slope in a hillside housing development to restore and to enhance primary visual elements: (*a*) on perimeter of development and (*b*) within development itself.

This approach is particularly well suited for restoring surface-mined areas to create outcroppings on man-made fill slopes, or other graded terrain, when such outcrops existed in the area prior to grading or when rock outcroppings occurred naturally in the surrounding terrain. In some mining-reclamation areas, rocks have been scattered at random as groups or individuals over the mined and "reclaimed" land, more as a disposal method of excess material. More recently these rock groupings have been designated by some agencies as "rodent condominiums" in the hope of encouraging the return of animal life. This practice leaves something to be desired as a single-purpose use. Rocks can serve more than one purpose if properly arranged and placed according to natural analogs!

In a natural landscape, rock can be found as caps on mesa formations, on side slopes as rock ledges and outcroppings, and along water courses as armoring. In all these locations, these elements not only serve a natural and aesthetic purpose but also provide erosion control (particularly in stream beds) and homes to assorted wildlife.

The use and placement of natural rock has been made recently in the surface-mining reclamation efforts of the San Juan and La Plata Mines in northern New Mexico, as illustrated in Figures 9.15 and 9.16.

9.12   COST CONSIDERATIONS AND ANALYSES

9.12.1   Land-Planning Costs

No additional cost should be experienced in developing plans that incorporate the landform grading concept. On the contrary, the creative opportunities and greater freedom this approach provides should motivate land planners to come up with cost-effective designs.

9.12.2   Design Engineering Costs

Design engineering costs may be higher if they are performed initially by a team only experienced in conventional practice. Initial grading-plan design costs by civil engineers who are inexperienced in landforming can increase from 10 to 15 percent. These costs decrease significantly once an engineer becomes familiar with the process. For example, after an initial encounter and experience, design costs can quickly decrease to a factor of 1 to 3 percent.

9.12.3   Surveying Costs

Similar cost considerations apply to surveying and field-staking control. With greater familiarity and experience, surveying costs decrease to 1 to 5 percent over conventional methods and approaches. Higher costs were initially experienced in the area of 10 to 15 percent with untrained crews. The survey team believed it was necessary to provide very detailed staking control in

Figure 9.15   Rock placed as ledges and outcroppings as part of a mining-reclamation effort to restore to preexisting conditions.

Figure 9.16   Rock placed as a cap on graded mesa as part of mining reclamation to restore to preexisting conditions.

particular for irregular breaks in slope. Each concave and convex shape tended to be covered with stakes. Sometimes this was a result of an inexperienced grading contractor's grade checker requesting a high density of stakes to provide a greater level of comfort.

This unfortunate practice once resulted in both comical and dangerous field conditions at a project site when a major rainstorm occurred after such extensive staking had been implemented. The rainstorm washed this artificial "stake forest" into the temporary erosion-control devices—the detention and de-silting basins—where the stakes all congregated around the water inlet structures, clogging the openings and causing a temporary overflow.

Incidents like this one called for additional training and clarification of objectives and procedures. Eventually, it was understood by *all* parties that the intent of staking and ground control was to convey overall ground shape or slope form rather than its precise mathematical configuration. Consequently, with a trained and experienced grade checker and properly prepared equipment operators, survey control of grading-operations costs can be brought back in line with conventional grade-staking costs. Today, more frequent usage of the GPS system has further simplified grading controls.

9.12.4   Landscape Architect Costs

Past experience has shown that landform designs impose little or no additional costs for the work of the landscape architect. A rethinking of conventional approaches to planting designs by the owners or principals of a landscape architecture company and a retraining of their staff may be all that is required. Reliance can no longer be placed on uniform coverage and repetitive plant patterns as a guiding principle.

9.12.5   Construction and Grading Costs

Construction and landform-grading costs are most directly related to the size and volume of earth movement than any other factor. Volumes of 1 million cubic yards or more historically experienced a cost increase of less than 0.5 percent. Important factors controlling grading-cost increases include the following:

- *Competitiveness of the market place.* Competition for larger projects, such as those for 1 million cubic yards or more, tends to eliminate landform grading standards as a significant factor.
- *Experience of the contractor.* Level of experience and coordination between the civil engineering consultants, the contractors on the job, and developer's representative.
- *Cooperation and understanding of the inspecting agency or agencies.* Approving agencies must be brought into the information-dissemination

process so that plan check, permitting and inspection can proceed smoothly. All the above will translate later into the all important prompt payment of billing to the contractor.

Grading costs in hillsides of largely sedimentary materials that require no blasting or extremely heavy ripping range from $2 to $5 per cubic yard with an average of $2 to $4 per cubic yard (in 2006). But, even between southern and northern California, significant differences in costs exist. Variables affecting the unit cost include the quantity of material, the nature of the operating area (open or confined), the earthwork logistics (i.e., length and steepness of the haul from the cut areas to the fill areas), and whether it can be ripped by conventional dozer and hauled by scraper equipment.

At first glance, it may appear that landform graded projects would be significantly more expensive to construct than conventional ones because of the grading and finishing of the more intricate details and the natural shapes required. However, experience has shown that the differential is minor when compared to the total project cost. This is true because the largest percentage (on average about 90%) of the earth volume moved, the mass "X" (shown in Figure 9.17) can be moved, placed, and compacted in a totally conventional manner. Only the outer slope layers, 20–50 feet thick (or approximately 10% of the total volume), require specialized shaping. Moreover, even this outer layer can still be placed and compacted with conventional equipment and methods. This outer component does need an additional grade checker for control and an experienced dozer operator for final shaping. The only significant and more intense additional field work required is the final finishing of the outer slope face features.

Accordingly, when costs are reckoned on the basis of the actual additional operations involved, they are a minor component, typically on the order of 1 to 2 percent of the total cost. Experience has shown that with in-depth training of the grade checker and equipment operators and other elements being equal (i.e., logistics) landform grading can be performed fairly routinely within the same basic time frame as traditional grading.

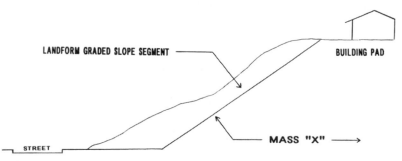

Figure 9.17 Relative amounts and location of earth movement by conventional as opposed to landform grading.

10 Public and Regulatory Response to Landform Grading

10.1 INTRODUCTION

This chapter discusses public acceptance of the practice of landform grading. What are the perspectives of the various players in the development process, namely, developers, planners, civil engineers, and regulatory agencies? What are the barriers to the acceptance or incorporation of new approaches into existing standards? What are some of the specific benefits of incorporation? Finally, what does the future hold for landform grading and what new applications lie on the horizon?

10.2 THE DEVELOPMENT PROCESS

10.2.1 Overall Governing Agency or Authority

Governing agencies—whether city, county, or state—have various provisions or procedures at their disposal to manage the development process and to guide and to control hillside development. Each one of these provisions has been prepared by and is under the jurisdiction of two very different entities, planning and zoning on one hand and engineering or public works on the other. These two professions, and their promulgating agencies, by their nature and training represent very different expertise and interests.

10.2.2 The Land-Planner's Perspective

One could say that land planners focus on land use, density considerations, and open-space preservation. Planners generally attempt to ensure that development proceeds in an orderly fashion that respects environmental constraints and community values. Principal documents and regulations that are directed at planners and developers include: (1) the General Plan, (2) the Specific Plan, (3) the Zoning Ordinance, and (4) the Hillside Management Ordinance.

Developers typically take their proposal for a project to a planning firm to determine what kind of land uses, densities, and yields they might obtain for a given parcel of land. The planner will develop various circulation patterns and create land-use "bubbles" and project densities. Often planners by background and training are not as well versed in viewing and interpreting a hillside three-dimensionally from the contour lines of a topographic map. Some people have an innate and intuitive talent for this ability. Sometimes it can be learned through extensive training. So while a planner's plan may have all the desirable aesthetic and marketing elements, it still has to be tested for practicality, engineering feasibility, and constructability.

After planners have drawn up their general and specific plans, the next step is to have a civil-engineering consultant review and critique the plans and determine their technical feasibility. Very often the development proposal goes back to the land planner with the engineer's critique and then back to the civil engineer until a compatible plan has emerged, one that the developer will feel comfortable about presenting to a public regulatory agency.

10.2.3   The Civil Engineer's Perspective

In contrast to the planners, civil engineers focus their attention on practical issues and problem solving, for example, on the requirement that a grading design meet a city's grading standards and that constructed slopes are stable following the recommendations of the geotechnical engineer. The civil engineer makes sure that drainage facilities will safely protect the development project from storm events, that roads meet safe-travel design standards, and that adequate water, sewer, and other services will be provided. Furthermore, the civil engineer should check to make sure that all the proposed facilities are constructable.

The most important documents and regulations for civil engineering and its various subspecialties are the following: (1) the Grading Ordinance and (2) the Design Standards for roads, storm drains, sewers, and water supply. The civil engineer will need to answer the following questions: Will the proposed roads meet permissible, safe-road standards for grades and horizontal and vertical curves? Will the plan allow for proper disposal of storm drainage? Will the site plan provide grades and elevations that allow for gravity sewer flow (to avoid costly sewer-lift stations)? Are there sites provided for water reservoirs that can serve all the pressure zone requirements of the individual segments of the development proposal located at different elevations?

10.2.4   Regulatory Agencies' Perspectives

Typically a developer will begin the submittal and review process with the local planning department. Planners tend to set standards that are often very general in nature and spell out intent and desired broad outcomes. The guidelines they provide are often subject to a great deal of interpretation. Under such circumstances, the developer and his or her engineer will interpret the guidelines to favor the plan they are proposing.

A classic example comes to mind. Frequently, the hillside-management ordinance will call for items to improve the aesthetics of grading, such as "all transition zones between a cut slope and the adjoining natural topography shall be rounded to provide for a 'smooth' transition to better 'blend with the terrain.'" The planner most likely intended a significant, clearly noticeable blend, that is, a curve with a significant radius of some 50 or 100 feet or an irregular, radius-controlled transition. More often than not, for a civil engineer, this means to be "frugal" with the earth movement, and he or she will show on a grading plan a 5- or 10-foot radius, something barely noticeable in the field and especially on any higher slope.

Unfortunately, the public planner in most jurisdictions does not review the grading plan and likely does not see it. Processing, reviewing, checking, and approving remains entirely the province of the engineering department and as far as the latter group is concerned a 5- to 10-foot radius has complied with the rounding requirement.

That however was not at all what the planner had in mind. To the planner, rounding is based on aesthetic considerations, that is, it is intended to replicate a pleasing natural landform or to provide for a far more gradual transition from man-made to natural terrain.

10.2.5   Owners' and Developers' Perspectives

The attitudes and views of owners and developers of projects should also be taken into account as well. These parties have a financial stake in the progress and outcome of their projects. Holdups and revisions in development plans translate into financial losses. They have to deal with tedious, time consuming, and often frustrating challenges to their projects from regulatory agencies, community associations, and environmental groups. These regulations and challenges can produce the type of angry response captured in the photo shown in Figure 10.1.

One of the main challenges that developers face, particularly in urban, hillside developments is opposition to proposed grading plans. A significant advantage of adopting a landform grading and revegetation approach is that it can greatly mute the objections raised by parties opposed to a proposed development.

10.2.6   Interdisciplinary Team Approaches

This disconnect and lack of coordination between different disciplines and professions in the development process suggest the need to set up a mechanism for closer coordination and cooperation in large-scale hillside-development projects. One possibility is a team concept in which the planning and engineering disciplines work closely together. In other words, the private planner and hillside-grading designer or engineer should not only work in the same building but have their desks next to each other.

This team approach would require that college and university programs for planners include a course (or courses) that includes some training in the fundamentals of topographic maps and their interpretation. This entails being

Figure 10.1 Sign posted at rock quarry reclamation site protesting regulatory challenges and restrictions on project.

able to visualize and to interpret contour lines on a two-dimensional topographic map into a three-dimensional picture of the landscape. This three-dimensional picture, must then translate into implications for any proposed hillside development or grading plan.

10.3   STANDARDS AND CODES

10.3.1   Difficulties with Promulgation

A charge sometimes leveled against the practice of landform grading is the general lack of standards and codes governing its use. Designers often depend upon or lean heavily on existing standards. Codes and standards bring order and certainty to a world of chance and variety. What penalty do we pay, however, for unexamined standardization?

The following epigram offers an interesting response to this question:

> Standards are always out of date . . . that is what make them standards. (Bennett, 1969)

As the epigram from Alan Bennett's play deftly recognizes, it takes time for a consensus to develop that can be codified. By then, new thinking may be emerging; but the presence of existing standards exerts a chilling effect on the development and implementation of new approaches. This chilling or negative effect is particularly manifest in a litigious society where any soul who dares to try something new—that is, neither "state of the practice" nor prescribed by codes and standards—is likely to be sued in court if anything goes amiss.

Landforming by its very nature somewhat defies standardization and codification. Landforms can be categorized to some extent—an effort has been made to do so in this book—but many combinations and permutations are possible. Some judgment and latitude are still required in each particular situation. It is difficult to fix standards because of the myriad of different conditions that can exist on any given piece of hillside property. However, if a plan is reviewed by both a trained planner and an engineer, who is open to new ideas and approaches, then it should be possible to decide whether a plan meets its intended purpose. Furthermore, it should also be possible to determine whether a design can be executed. At the very least, this review should provide valuable input to a developer and his or her consultants.

10.3.2 "Prescriptive" vs. "Performance" Standards

Prescriptive and performance standards both have their advantages and limitations. Performance standards are preferable from a theoretical standpoint, because they allow for greater flexibility and innovation in both design and execution. Performance goals are specified, and then it is up to the designer to come up with a plan that meets these goals. This approach has limitations, though, as it assumes a pool of qualified design talent and requires careful inspection and monitoring by qualified and appropriately trained reviewers.

Prescriptive standards typically specify materials and method of execution. This approach is favored where there are limited funds and/or qualified personnel to conduct the necessary inspection, monitoring, and evaluation of proposed designs. Innovation and potentially superior design are sacrificed in favor of possibly inferior designs that at least meet minimum criteria with some degree of certainty.

10.4 PROJECT-APPROVAL BENEFITS OF LANDFORM GRADING

The 2000s continue to be a decade of concern for the environment. More than ever, the public and its government representatives are aware of the impact that development has on the natural landscape and demand sensitivity in its application.

In addition to the protection of fauna and flora, protection and/or restoration of significant visual assets have received increasing scrutiny. Hillsides are some of our most noticeable and precious visual assets. Mass grading, in the past, has radically altered that landscape without paying sufficient attention to the visual scars and discordant landscape elements that can be produced in the process.

As a result, mass-grading proposals have increasingly encountered strong opposition, and more stringent controls and limitations have been placed upon development proposals. Landform grading and revegetation provides a possible way around this dilemma. Hillside-development proposals that incorporate this approach are more likely to be approved because these designs are more attractive and more natural. Past experience has shown that specific benefits and advantages can be derived from the application of landform grading and revegetation:

- Projects that might have been rejected can receive approval. Examples of such projects are discussed in Chapter 12; they include Talega and Hollywood Hills.
- The approval process can be accelerated (e.g., Talega and Hollywood Hills).
- Less stringent stipulations may be applied in some areas while certain development standards might be relaxed in others (e.g., Anaheim Hills).
- In some cases, communities have granted density bonuses for the application of landform grading to offset any cost increases or loss of buildable pad areas (e.g., Los Angeles).
- Objections from neighboring, existing developments, open-space preservationists, environmentalists, and no-growth advocates can be limited by offering landform grading and revegetation and its accompanying benefits as trade-offs and mitigation measures (e.g., Anaheim Hills and Talega).

10.5 AGENCIES THAT HAVE ADOPTED OR IMPLEMENTED LANDFORM GRADING

Several different agencies have accepted or adopted the *Landform Grading* concept either with or without formalities. The following is a list of projects, relevant regulatory agency, and a brief description of how this adoption was managed. Many of these projects are described in the case studies in Chapter 12.

1. *Anaheim Hills Project, City of Anaheim, California*
   This project, approved in 1976, was the first permitted application of its type. The director of public works, city engineer, and planning director first approved the landform grading concept at staff level, allowing a test project to proceed. Upon completion of the test project, the staff conducted field-inspection tours for city council members and city planning commissioners. Upon acceptance by all, the City of Anaheim adopted a modification to their grading ordinance, which is still in effect.

2. *Talega Project, City Portion, City of San Clemente, California*
   The project used separate formal public hearings as part of the development process. Meetings were held first between the developer and his consultants with city staff, including the community development director and the city engineer. Then the proposed project was presented to the community at large in a public hearing setting with city staff present to observe the community's response; next, to the planning commission; and finally to a joint session consisting of the City's planning and engineering staff, the planning commission and city council.

   When interviewed more than a decade later, the director of the city of San Clemente's community development department is still a strong proponent of landform grading. He commented: "Our experience with landform grading was excellent. Landform grading along the primary entry corridor to this planned community not only extended the natural canyon topography into the project and beyond but also created premium values for the master developer and the individual home builder." The director has assembled a photographic record that is a side-by-side comparison of the "Talega Landform and Revegetation Experience" with a similar project in Orange County that was graded, under a different jurisdiction, using the conventional approach. He keeps this record handy in his office to illustrate to future city developers the dramatic differences between the two approaches.

3. *Talega Project County Portion, County of Orange, California*
   No public hearing was conducted but an internal decision was made by engineering and geotechnical departments of the county. The manager of development services for the county was very supportive of the landform grading approach and noted that the grading was accomplished in area that had very unstable soil conditions. He concluded that "the use of landform grading seemed to be a perfect blend of basic principles of nature and good engineering practice." Formal guidelines were drawn up and mutually agreed upon by the landform-grading consultant in cooperation with county staff and then adopted and passed on to plan-reviewing and inspection staff for use.

4. *Hollywood Hills, Department of Water and Power, City of Los Angeles, California*
   Approval for this project proceeded along two separate lines: (1) response to public hearings and community demands and (2) internal acceptance by engineering and geotechnical departments. The project was approved with design modifications in response to concerns raised by the public and participating city's engineering and geotechnical staff. Some of the modifications were implemented after the project started.

5. *Lake Asaayi Project, The Navajo Nation, Window Rock, New Mexico*
   The Navajo Nation Environmental Protection Agency, Water Quality Division, staff reviewed and studied the landform restoration concept in

competition with a conventional approach. After the staff recommended adoption, the tribal council voted unanimously for approval of the alternative landforming concept.

6. *Planning Department, City of Los Angeles, California*

Planning staff prepared a proposed modification to the city hillside-management ordinance after the senior author conducted field trips to ongoing landform grading projects in Anaheim Hills for city representatives and developers and their engineers in the Los Angeles area. After considerable discussion among interested parties, a document was adopted ("Planning Guidelines—Landform Grading Manual") that made the application of landform grading and revegetation optional rather than required. However, city regulations provided for density credits whenever a developer opted to use the new planning guidelines.

7. *Surface Mine Reclamation Projects, State of New Mexico*

Decisions to apply a landforming approach were made to improve current practices, to solve erosional problems associated with conventional techniques, and to reduce runoff control costs. Some reclamation projects have been designed using a computer software program that applies fluvial geomorphic principles to upland landform designs (Bugosh, 2006). The essential features or highlights of a landforming approach to surface-mining reclamation are shown in Figures 10.2, 10.3 and 10.4. An alternative landforming approach attempts to replicate the original, vertical bluff lines and rock ledges characteristic of natural terrain in the area as depicted in Figure 10.4. The landform shapes created reflect natural topographic analogs of the surrounding undisturbed terrain. Note how the topography is graded not to create a co-

Figure 10.2 Dozer at work reshaping tailing dump into landforms. The irregular mounds of earth on top of the ridgeline are stockpiles of topsoil to be applied to the slope face upon completion of landform shaping.

Figure 10.3 Completely landform-shaped tailing deposit that replicates the natural analog. Tailings deposit does not have any of the usual features of a conventional tailings "dump," namely, angular and linear forms, planar surfaces, terraces, and armored drainage ditches.

Figure 10.4 View from the top of a landformed tailing deposit, looking downhill. Notice the shaped, meandering flowline that is minimally armored with naturally interspersed rock ledges.

hesive mass but rather to segment it into smaller individual tributary drainage areas through the use of interspersed valleys and ridges that are shaping the main tailing body. The entire topography also has a characteristic, primary, natural ridge line with vertical variations.

Typical topography of reclaimed areas without landform restoration and the benefit of "approximate original-contour" reclamation is depicted in Figure 10.5. Figure 10.6 depicts the same area as in 10.5 but is redesigned with the help of the geomorphic and landforming principles and thereby recreating an "approximate original" contour condition complete with natural landforms and a reestablished natural hydrologic pattern.

Alternatives to the present "high-wall" elimination approach in surface-mining reclamation are depicted in Figure 10.7. An alternative way of handling concentrated runoff is shown in Fig. 10.8.

8. *Rosebud Mine, Western Energy Company, Colstrip, Montana*

This project involved the reclamation of a major, residual mining feature, a former coal seam trench and the corresponding high wall. The height from the bottom of the pit to the top of the high wall was about 120 feet. Landforming techniques were applied to fill in the trench and then to fill and shape landforms against the high wall, replicating the natural, undisturbed topography adjacent to the site, with a series of ridges and valleys. At the same time, this topography created smaller individual tributary drainage areas, avoiding the typical rill erosion of a linear, planar slope. Views of the coal-mine trench before and after

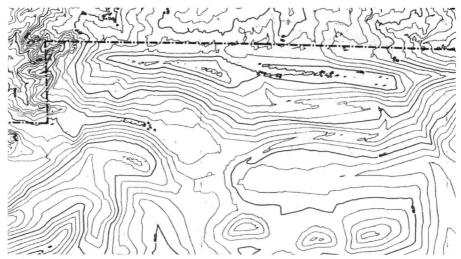

Figure 10.5 Reclamation plan of mined area without application of "approximate original contour" (AOC) concept.

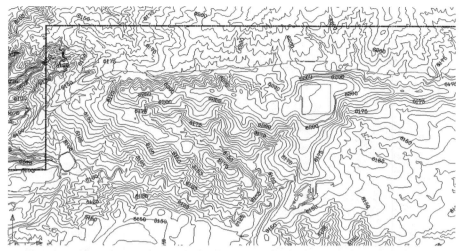

Figure 10.6 Reclamation plan of same mined area as shown in Figure 10.5 after applying Geofluv™ computer program that replicates characteristics of the original underlying topography (adapted from Bugosh, 2006).

Figure 10.7 Natural high-wall features preserved by placement of partial, concave-shaped fill slope against vertical bluff.

Figure 10.8   Interim drainage-collection pond prior to release into natural main channel.

reclamation, using landforming concepts, are shown in Figures 10.9 and 10.10. This is a classic example of how the essential features of a high wall can be preserved without cutting back and further damaging natural topography.

On the other hand Figure 10.11 illustrates how unnatural and destructive the strict application of an East Coast developed OSM standard can be in a totally different environment. Notice how in this New Mexico setting the two original highwalls on the right and left were cut back while a pre-existing vertical bluff (or highwall) can be seen in between. The latter provides a natural, on-site analog that is ready to be replicated. Instead this natural analog was ignored and a foreign landform was forced upon this landscape

In this case the mine owner believed that several conditions warranted this approach rather than the usual practice of cutting back the high wall at safe angles. Fortunately, Montana's Department of Environmental Quality Administrative Rules of the State provided for an exception to their standardhigh-wall reduction regulation if the alternative proposal will replace bluff features that existed before mining and the department determines that:

1. Postmining bluffs are stable with a minimum long-term safety factor of 1.3.

2. Postmining bluffs will be compatible with postmining use.

3. Similar geometry, function, and horizontal linear extent of the new bluffs do not exceed premining conditions.

Figure 10.9   View of exposed 120-foot deep trench prior to start of mitigation work.

Figure 10.10   View of the same trench area totally reclaimed and topographically restored using landforming techniques.

Figure 10.11 Photo illustrating the scarring practice of laying back mined highwalls in a landscape where "highwalls" (vertical bluffs) are naturally occuring landforms.

Additionally, in this case, the alternative approach also saved some 5,000 pine trees; the high-wall reduction-fill slope faced north and thus and helped with the re-establishment of plant growth. The retained trees from the preserved bluff also provided an excellent long-term seed source. The department soon recognized the benefits of the alternative proposal and, according to the Western Energy Company, was supportive from the start.

10.6 FUTURE APPLICATIONS OF LANDFORM GRADING

Landform grading is not limited only to hillside development projects, mass-grading applications, and watershed-restoration and mining-reclamation projects. This approach is applicable to virtually any hillside grading that results in highly visible slopes and landform and where visual or aesthetic impacts are of primary importance. It should be stressed that a wide variety of landforms including valleys, arroyos, and stream elements are amenable to a landform grading approach;em- not simply hillsides and slope elements. The Asaayi Lake Watershed Restoration Project (see Chapter 11) is a good case in point. Here hillsides were not an issue; the primary goal was to re-establish a stable natural wash or valley.

Possible candidates for landform grading include slopes along major highways, collectors, and other thoroughfares, as well as other highly visible and permanently exposed areas and slopes along parks, open spaces, golf courses, etc. It need not be applied, on the other hand, to minor slopes, slopes below roadway grades, side slopes between homes and other structures, minor rear-yard slopes. and any other slopes that would ultimately be obscured from view.

The Landform Grading and Revegetation concept has yet to make significant contributions in the following areas that have had a major negative environmental and visual impact on the natural landscape:

- Highway and freeway construction
- Above ground sanitary landfills (artificial, ubiquitous, and highly visible types of urban landforms)
- Earthen dams (they too are landforms that significantly impact valley landforms)

These areas are still characterized to a large extent by conventional mass grading in which the original topography has been totally remanufactured, thereby creating artificial, alien landforms with linear, planar, uniform surfaces.

In contrast, landform grading can create an artificial hillside designed according to the principles and techniques described in Chapters 7 and 8 that

Figure 10.12 Artificial hillside designed according to landform grading principles consisting of a 75-foot high fill (foreground) that is in complete harmony with the natural, undisturbed hillside (background).

is in complete harmony with surrounding undisturbed, natural terrain as shown in Figure 10.12. Notice the similarity of the shadow patterns between the natural and man-made slopes that accentuate the concave and convex slopes.

10.7 REFERENCES

Bennett, A. 1969. *Forty Years On.* London: Faber.

Bugosh, N. 2006. *Primer to the GeoFluv™ Method for Reclamation Landform Design. Paper* (unpublished) presented at the National Interactive Forum on Geomorphic Reclamation. Sponsored by the Office of Surface Mining, Western Region, Farmington, N.M., Sept 12-14, 2006.

11 Landforming Projects— Watershed Restoration and Mining Reclamation

11.1 INTRODUCTION

The next two chapters describe actual projects where *landforming* techniques were employed. This chapter describes projects where landform grading was used for mining reclamation and watershed restoration. These projects include: (1) runoff and erosion control, School Girl's Glen, University of Michigan Arboretum, Ann Arbor, Michigan; (2) watershed restoration, Navajo Nation, New Mexico; and (3) mine reclamation, Syncrude Canada oil-sands project, Alberta, Canada.

An attempt has been made to adopt a common format to describe each case-study project, including the following elements:

- Project Name and Location
- Project Type
- Client or Project Owner
- Site Conditions and Problems
- Treatment Considerations
- Selected Treatment
- Performance Evaluation
- References

These case studies were selected to demonstrate the principles and guidelines discussed in previous chapters of the book. Each case study is illustrated liberally with photos and diagrams.

11.2 SCHOOL GIRL'S GLEN

11.2.1 Project Type

The goal of this project was erosion and runoff control as part of watershed restoration.

11.2.2   Project Location

School Girl's Glen is a steep ravine in the Nichols Arboretum in Ann Arbor, Michigan. The arboretum, which adjoins the University of Michigan to the east, is located on the flank of a glacial moraine that overlooks the Huron River Valley. The glen is part of the Huron River drainage, which flows into Lake Erie. The stream that flows out of the glen is intermittent and drains a watershed of approximately 33 acres. The stream is fed entirely by storm-water runoff, so it fluctuates between being nearly dry and discharging large amounts of water.

11.2.3   Client and Project Owner

Nichols Arboretum is administered and maintained by the University of Michigan, Ann Arbor.

11.2.4   Site Conditions and Problems

The problems facing School Girl's Glen are multifaceted. High runoff from impervious surfaces in the watershed, combined with direct discharges from street drains, have resulted in large storm flows and serious erosion problems. Concentrated runoff from roads in a cemetery above the glen flowed unchecked past the arboretum's environmental interpretive center, which made access to the building difficult during storms and completely disrupted plans to construct a planned "gateway" garden at the west entrance to the arboretum.

Uncontrolled storm flows had caused both channel head-cutting and down-cutting in the glen itself. This, in turn, has undermined adjacent stream banks and side slopes, which has resulted in serious slope failures (see Figure 11.1). The combination of surficial erosion, channel scour, and slope failures produced massive soil losses. At low water a large sediment delta was clearly visible at the confluence of the glen's ravine and the Huron River.

A cavity analysis using topographic maps (see Figure 11.2) from 1965 and 1993 estimated that 13,500 cubic yards of soil washed out of School Girl's Glen and into the Huron River during this time period. This estimate works out to an average of about 480 cubic yards per year or an average of about 1.3 cubic yards of soil per day washed out of the glen. Given the fact that the glen has been handling an ever-increasing amount of water, this soil loss has probably accelerated during the last decade prior to repair and restoration work.

The soils in the glen are well drained to moderately well drained sandy loam to clay loam over a hard till, which is exposed at the bottom of the ravine. Soils exposed in side-slope failures are variable, with boulders often embedded within the matrix. The areas with sandy to loamy soils are highly erodible when the vegetation is removed. The hard-packed clay till, which

Figure 11.1   Slope failure in School Girl's Glen caused by stream scour and under-mining.

lines parts of the ravine is less erodible, but it is still vulnerable to down-cutting during storm flows.

11.2.5   Repair and Restoration Goals

The main goal of the School Girl's Glen watershed repair and restoration project was to arrest the erosion and slope failures occurring in the glen and to eliminate sediment discharges to the Huron River. A related goal was to carry out this remedial work as a demonstration project using environmentally sensitive stream-bank protection and erosion-control methods. A highly desirable concomitant goal would have been to work on source reduction—i.e., to reduce or divert the amount of runoff entering the glen. For a variety of reasons this was not an option available to the project.

11.2.6   Treatment Considerations

Work began in School Girl's Glen in July 2001 and continued through August 2003. A variety of different techniques were implemented in School Girl's Glen, to correct and mitigate existing problems. Although source reduction was not a project option, some effort was directed at detaining and controlling runoff entering the head of the glen. This was accomplished mainly by con-

Figure 11.2   Topographic map of School Girl's Glen area showing general relief and margins of the ravine.

structing a series of landscaped, stepped pools. Other measures were implemented to better handle street-drain discharges that had initiated tributary gullies on the side slopes of the glen.

The techniques that were considered were relatively low-cost, environmentally sensitive measures that could be implemented using volunteers or arboretum staff labor.

11.2.7 Selected Treatments

A range of techniques were selected that met the preceding requirements—low cost, environmentally sensitive, and using arboretum staff labor. These treatments included the following:

- Stone weirs
- Live staking and native planting
- Gabion check dam and root-wad revetment
- Invasive plant and tree removal and revegetation
- Reinforced earth–repair fill
- Boulder cascades
- Stepped pools

These treatments are described in detail elsewhere (Gray et al., 2004). A brief description of one of these techniques that has an important grading component, namely, stepped pools, is described below.

A series of stepped pools were constructed at the head of School Girl's Glen (see Figure 11.3) to handle and slow runoff flowing into the head of School Girl's Glen. Generally, step-pool structures occur in natural, steep gradient, perennial streams. The pool system that was built in the arboretum is perhaps better described as a series of low-height, boulder drop–structures connected by a low-gradient channel (see Figure 11.4). These structures lower existing channel velocities and provide a stable channel environment. A total of six pools or low-height boulder drops were constructed. The top three pools handle runoff from Forest Hills cemetery, which had previously flowed unchecked past the environmental interpretive center building (see Figure 11.5).

The runoff flow into these pools originates from a subwatershed that has a 17-acre drainage area, consisting mainly of cemetery property that contains approximately 6,600 feet of 11-foot-wide asphalt and compacted gravel roads. The calculated peak 10- and 100-year-recurrence-interval flow rates for these pools are 5.0 cubic feet per second (cfs) and 16.5 cfs, respectively. An outlet drain conveys additional storm runoff from an adjacent subwatershed with a drainage area of approximately 1.3 acres. This drainage area includes mostly impervious roadway and roof surfaces and some surrounding lawn areas. The calculated 10- and 100-year-recurrence-interval flow rates from the combined

Figure 11.3 Stepped pools under construction in front of arboretum environmental interpretive center.

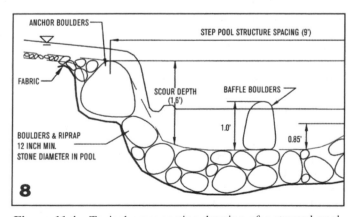

Figure 11.4 Typical cross-section drawing of a stepped pool.

(a)   (b)

Figure 11.5   Runoff conditions in front of environmental interpretive center building: (a) uncontrolled and (b) after construction of stepped pools.

flows that must be handled by the lowermost three pools are 7.0 cfs and 19.7 cfs, respectively.

The channel between pools was designed to carry the 100-year peak flow at a nonerosive velocity. A triangular channel at a 1-percent slope was selected to minimize earthwork and provide a subcritical channel design. Based on Manning's equation, the maximum velocity in the channel during the 10-year event will be 2.6 feet per second (fps). It was assumed that a minimum of Type B vegetation (NRCS, National Resources Conservation Service) would be required to adequately stabilize the proposed channel. Type B includes various native Midwest grasses.

A variety of prairie and river's edge species—such as riverbank sedge (*Carex emoryi*), prairie cordgrass (*Spartina pectinata*), Gray's sedge (*Carex grayii*), Indian hemp (*Apocynum cannibinum*), blue flag iris (*Iris virginica*), and cupplant (*Silphium perfoliatum*)—were planted in the channel. Calculations showed that the allowable velocity in a channel with this type of vegetative lining is 6.6 fps. Accordingly, a Type B vegetative cover will provide adequate stabilization in the channel.

Construction of the pools and connecting channel commenced and was completed in the summer of 2003. A considerable amount of earthwork grading was required to shape the channel and surrounding banks. A geotextile-filter fabric was placed behind and beneath the boulders to prevent washing of fines through the rock. Topsoil was added to the regraded surfaces and then covered with a straw-filled erosion-control blanket that was securely fastened to the ground (see Figure 11.6). A selected spectrum of grasses, herbaceous vegetation, and shrubs were then transplanted or sprigged through the blanket into the topsoil beneath.

11.2.8 Performance Evaluation

The stepped-pool system has been tested severely by several storm events that occurred shortly after construction (see Figure 11.7) and before establishment of vegetation. Nevertheless, the system has performed well and largely as expected. The pools temporarily detain storm runoff and even out the flow of water. Temporary detention in the ponds, which have pervious bottoms and native vegetation, also permits some of the runoff to infiltrate the ground, as opposed to discharging into the head of the glen below. Phreatic vegetation (sedges, rushes, etc.) growing in the pools and connecting channels help to clean runoff water as it flows through them in a process called phytoremediation.

The stepped pools and channels are now well vegetated and attractively landscaped (see Figure 11.8). The pools and landscaping greatly enhance the entrance into the visitor center and arboretum grounds. Runoff flowing past

Figure 11.6 Straw-filled erosion control blanket placed over topsoil cover.

Figure 11.7   Stepped pools after a major rainstorm successfully conveying runoff.

(*a*)

Figure 11.8   View of stepped-pools and channel in Nichols Arboretum two years after revegetation: (*a*) Lower pools and (*b*) upper pools.

(b)

Figure 11.8   (*Continued*)

the visitor center, which hitherto had been a liability, has been converted into an asset.

11.2.9   References Cited

Gray, D. H., R. Grese, and T. Orlow. 2004. Saving School Girl's Glen: Erosion control and watershed restoration in a university arboretum. *Erosion Control* 11 (2):34–45.

11.3   ASAAYI LAKE NORTHWEST DRAINAGE—LANDFORM RESTORATION

11.3.1   Type of Project

Watershed restoration to stop valley erosion and siltation of a lake on Navajo tribal lands is this project's goal.

11.3.2   Location

The site is located in the Chuska Mountains in New Mexico on the Navajo Reservation at an elevation of 7,500 feet, some 60 miles north of Gallup, New

Mexico, the nearest largest city, or approximately 15 miles northeast of Window Rock, the capital of the Navajo Nation government. Asaayi Lake is also known as Bowl Lake on United States Geological Survey (USGS) maps.

11.3.3   Client

The client for this project was the Navajo Nation Environmental Protection Agency's Water Quality Division, located in Window Rock, New Mexico.

11.3.4   Repair and Restoration Goals

The goals for this project were to "heal" the damaged valley by reshaping its landform through geomorphic alteration so that it replicates the Southwestern United States's "wash" landscape, which is less prone to erosion, and consequently to stop significant lake siltation.

11.3.5   Site Conditions and Problems

The Asaayi Lake project is a classic example of how *Landform Grading* principles lend themselves to a variety of applications, not only to slope forms and shapes but even more so to entire landscapes of different topographic forms. The site in question was affected by serious erosion problems.

A deeply eroded valley, over 1,400-feet long, discharged sediment into a water storage and recreational lake at its terminus. The tributary drainage area of the Northwest Drainage to the lake was over 200 acres. The overall original natural grade in the valley was 6 percent. The total runoff flowing down the valley was approximately 800 cfs (based on a 25-year-storm condition).

In 1968 the U.S. Bureau of Reclamation had constructed a dam and a reservoir downstream from the valley to provide the tribe with two primary opportunities, namely, water supply for agricultural irrigation and recreational (fishing and boating) use. The addition of a campground and community-gathering facilities further enhanced the public use of the site. A photo of the lake and campground are shown in Figure 11.9.

This valley has had a history of aggradation and degradation over long periods of time. The latest period, one of erosion, occurred in the deep alluvial deposits throughout the valley. The main degradation or incision happened along the center of the valley, along its north-south axis. Deep erosion channels also developed on the west side, from runoff entering the valley from the adjacent mountains. Much of the erosion could be attributed to the construction of a logging road, which later became the primary access road to the lake and recreational facilities. As roads often do, this diverted and concentrated runoff in unfavorable locations. Uncontrolled seasonal grazing, with its attendant trampling and destruction of ground cover, also exacerbated erosion problems. Photos of deep gullying and channel incision along the main valley and its tributaries are shown in Figures 11.10–11.12.

Figure 11.9 Photo of Lake Asaayi showing lake in background and campsite in the foreground.

The result of this erosion was not only a deeply incised valley floor, up to 40 feet deep (see Figures 11.10 and 11.11), but also a deposition of sediments downstream into Lake Asaayi. This deposition reduced the holding capacity of the lake to the point where it was beginning to impair all its intended functions. By the year 2000, the lake had lost substantial amount of its holding capacity. Action was required to prevent the ultimate siltation of this body of water and possible dam failure.

11.3.6 Treatment Considerations

The tribe initially engaged the services of the Natural Resources Conservation Service (NRCS). The agency engineers developed a detailed plan to mitigate these problems, which called for the construction of a series of five large check dams (or flood retarding structures, as they were identified on the plans) to be built upstream in the tributary basin to intercept runoff, reduce flow velocities, and trap sediment. The agency earmarked funds in the amount of $2 million for the project.

This plan although already fully engineered and ready for construction (see Figure 11.13*a*) was ultimately rejected by the Navajo EPA, Water Quality Division, and the Navajo Tribal Council. These agencies determined that this

Figure 11.10   Photo showing severe erosion and gullying of main wash upstream of Lake Asaayi, view looking up valley.

plan was too invasive and damaging to the natural beauty of the forested mountain terrain. The construction of the five structures and their related equipment access roads would have damaged the forested tributary-drainage areas. Furthermore, it would have left the valley in its eroded state. This outcome was not acceptable to the tribe from either an aesthetic or recreational-use standpoint, nor did the NCRS plan allow any future use of the valley (e.g., for seasonal grazing).

11.3.7   Selected Treatment

Following hearings and additional study sessions, the tribe accepted an alternative landform grading and restoration solution proposed by the principal author. This approach proposed the creation of a broad and shallow desert wash characteristic of the southwestern landscape to replace the deep valley incisions. The plan was to achieve this through a complete regrading of the site and the installation of a series of grade-reduction structures or ramps constructed from rock.

The main axis of the eroded valley would be reconfigured with its deepest and widest gullies leading into a wash 100 feet in width at its upstream head and expanding to 275 feet in width at its downstream terminus. The reconfigured wash depth ranged from 10 to 20 feet. Its side banks were laid back to slopes, with gradients ranging from $1\frac{1}{2}:1$ to $3:1$ (H:V). Tributary side

Figure 11.11   Photo showing depth of erosion and gullying of main wash upstream of Lake Asaayi.

channels were to be similarly regraded at their downstream outlet or confluence with the main channel by flaring them out in the direction of the flow. A plan view of the reconfigured wash and tributary valleys is shown in Figure 11.13*b*.

This alternative plan was reviewed in detail by various departments of the NRCS, including the Fluvial Geomorphology Section in Fort Worth, Texas; the National Landscape Architect in Washington, D.C.; and the Civil Engineering Office in Phoenix, Arizona. All approved the design concept; plans for it were executed by the NRCS Civil Engineering Office in cooperation with consultation by the principal author of the plan. The engineer's estimate for the project and the approved funding were determined by the NRCS to be $1.5 million.

11.3.7.1   Initial Design Concept: The new concept called for a series of regrading and landforming treatments, which included the following elements:

1. Regrading of the eroded valley to the specified "wash" configuration (refer to Figure 11.13*b*).

Figure 11.12   Photo showing severe erosion and gullying in tributary valley upstream of Lake Asaayi.

2. The installation of approximately 10 rock-grade reduction and erosion-control structures (rock "ramps") on both the main channel and some tributary channels to reduce flow gradients to between 0.5 and 1 percent.

3. The preservation, where possible, of any existing trees or shrubs in areas to be graded.

4. The hydromulching of all grade areas with native-seed mixes consisting primarily of Indian Rice Grass and Western Wheat Grass.

5. Finally, the site was to be fenced with barbed-wire ranch fencing to prevent the premature entry of livestock prior to establishment of a durable vegetative ground cover.

11.3.7.2   Final Design Configuration: Before the design could be finalized, a drastic reduction in available funds caused a reevaluation of the project situation. The limitation of funds required that the most expensive component, the installation of the rock-ramp elements would have to be curtailed. The final, constructed design would have the following characteristics:

1. The central valley would still be regraded to remove the erosional scars and to create the broad natural analog "wash" configuration, as illus-

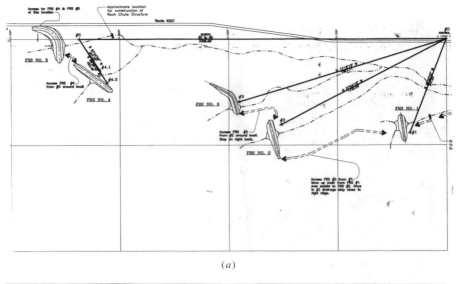

(*a*)

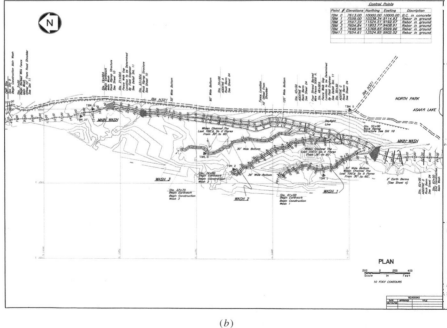

(*b*)

Figure 11.13 (*a*) Original plan showing location of proposed check dams in tributary gullies and new access roads (from USDA/NRCS, 2002). (*b*) Plan-view drawing of reconfigured main wash and tributary valleys (from USDA/NRCS, 2002).

trated in Figure 11.14. This wash was kept within the original width specifications that ranged from 100 to 275 feet.

2. Rock-ramp grade-reduction structures had to be limited to three on the main channel—one at the head scarp upstream, one at midpoint (see Figures 11.15 and 11.16), and one at the downstream discharge point. Because of the now-limited erosion control structure, the overall grade would only be reduced to from 1 to 4 percent between reaches.

3. Three of the primary tributary side channels were regraded at their starting points, at their upper ends into broader-base washes and significantly flared out at their discharge points into the main channel. A rock ramp was placed at the mouth of the largest channel, where it discharged into the main channel. The fourth—uppermost and smallest—tributary channel was not altered because of its very narrow valley and forested rim. Its outlet into the main channel also was not rock treated, but instead a diagonal berm was placed across its mouth to create a small basin, as shown in Figure 11.17. This allowed ponding and stilling of the upstream runoff from the tributary before discharging into the main channel.

4. Existing trees and shrubs in the main channels and subchannels would, where possible, be preserved and grading conducted around them.

5. Funding for revegetation and fencing would have to be provided for by the tribal agency.

Figure 11.14 Photo showing severely eroded main valley being regraded into a broad, wash configuration.

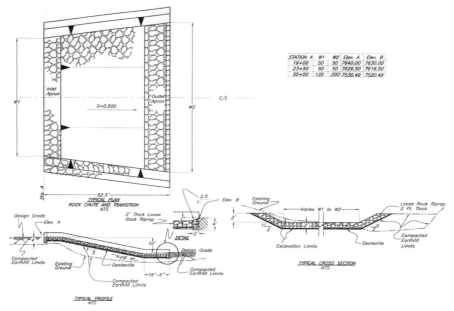

STATION A	W1	W2	Elev. A	Elev. B
19+00	50	50	7640.00	7630.00
23+50	50	50	7626.50	7616.50
55+00	120	200	7530.49	7520.49

Figure 11.15   Plan, profile, and cross-section views of rock-ramp structures (from USDA, 2002).

11.3.8   Performance Evaluation

The project was placed out for open, public bidding, based on these new design parameters. A local private contractor was selected who completed the construction in three months. Straw was applied on top of the hydromulching to provide additional interim protection prior to ground-cover establishment. Lack of precipitation, and later high winds, blew much of the straw cover away, preventing early germination. In spite of these setbacks, which also included funding reductions, relaxed specifications, and the aforementioned unfavorable weather conditions, the project has fared exceptionally well. The most important goal of all, namely, eliminating the destructive erosion in the valley while maintaining the natural character of the region, was achieved.

The erosion scars have healed, siltation of the lake has stopped, and the valley is beginning to blend naturally into the surrounding landscape. Views of the broadened, regraded main valley are shown in Figures 11.18–11.20. After the first year, only minor gullying in the channel bottom and side slopes had occurred. This was to be expected, because a substantial vegetative cover had not yet become established and residual gradients on the project of up to 4 percent were still in the erosive range. To counteract this localized erosion, the Navajo Water Division has placed straw bales across critical locations in the main channel (see Figure 11.18) to help reduce flow velocities until vegetation becomes better established.

Figure 11.16 Rock-ramp or grade-control structure located in main valley at its mid-point. Three other similar rock ramps were installed in the regraded main valley.

Figure 11.17 Tributary gulley regraded and flared out where it joins the main valley.

Figure 11.18   View looking down regraded and broadened main valley. Straw bales were placed across wash at critical locations to minimize erosion.

Figure 11.19   Appearance of valley three years after treatment, view looking downstream. Wash bed is stable, and erosional scars are completely healed. Photo taken September 2006.

Figure 11.20 View of valley, looking across the wash, three years after treatment. Vegetative ground cover is well established, and erosion has been arrested. Photo taken September 2006.

At the dedication ceremony in September of 2004, tribal, federal, state, and local government officials expressed their pleasure with the results of this landform restoration project and hoped that it would become a demonstration case for many similar sites existing throughout the Southwest.

11.3.9   Postscript

An inspection of the project site three years later revealed a stable, natural-wash bed, showing little or no signs of erosion and blending seamlessly into the surrounding landforms and landscape (see Figures 11.19 and 11.20).

In summary, the project was a success: Landforming treatment has returned the valley to its natural, pristine appearance, reestablished its original geomorphology, maintained its hydrologic integrity, restored the land for recreational use, both aesthetically and functionally, while at the same time eliminating deposition of sediment from the valley into the downstream lake. All this was accomplished with minimal intrusion of man-made structures and features. As a result, the Navajo agencies consider this project a model restoration for future, similarly eroded valleys on reservation lands.

11.3.10   References Cited

United States Department of Agriculture (USDA). 2002. *Asaayi Lake Northwest Drainage Channel Shaping and Grade Stabilization.* Design documents prepared for the

Navajo Nation and Fort Defiance Southwest Conservation District (SWCD). Natural Resources Conservation Service (NRCS).

11.4   OIL SANDS MINING RECLAMATION, SYNCRUDE CANADA

11.4.1   Project Type

The Syncrude Canada project is one of surface-mining reclamation and landscape restoration.

11.4.2   Project Location

The project is located in northern Canada, about 500 kilometers north of Edmonton, Alberta, near the town of Fort McMurray, on the Athabasca River. The project is also known as the Mildred Lake Operations.

11.4.3   Client

The client for this project is Syncrude Canada, Ltd., which is located in Fort McMurray, Alberta, Canada.

11.4.4   Site History

In the early 1920s, private interests first explored and extracted oil from the deeper oil sands deposits in a small scale operation along the Athabasca River; but this venture was not commercially viable with the equipment then available and the logistics problems encountered in this remote location. In the 1960's areas of northern Alberta owned by the Provincial Government that had been divided up in leases and subsequently leased were now being developed for the purpose of extraction oil from the oil sands deposits. Over the years, companies have explored different methods to extract the oil from this area. There are currently two main methods: surface mining and steam assisted gravity drainage (SAGD). Since the late 1970's Syncrude has expanded their operation and has become one of the largest leaseholders and mine operator and is today one of the world's largest producer of synthetic crude oil derived from oil sands. Syncrude's existing holdings lends themselves to surface mining only.

11.4.5   Site Conditions and Potential Problems

The landscape in this region has little topographic relief; it is heavily vegetated; and it has extensive rivers and meandering streams that exhibit extreme sinuosity, as shown in Figure 11.21. Some lakes and occasional large sand-dune fields also occur in several areas.

Figure 11.21 Aerial view of terrain illustrating low-relief, vegetative cover and stream meanders.

The oil-sands layer itself ranges from 75 to 100 meters in thickness and is covered by overburden of up to 70 meters thick. The excavation—which originally was made by a variety of methods, including bucket-wheel excavators and draglines—is now performed exclusively with mining shovels and truck haulers, as shown in Figure 11.22.

To illustrate the magnitude of the operations, at the Mildred Lake site, as of the end of 2005, a total of 17,037 hectares of land have been disturbed to date for mining related activities and about 4,296 hectares have been reclaimed while at Syncrude's Aurora operation, which began around 1998, as of the end of 2005, 4,167 hectares have been disturbed and 63 hectares reclaimed. Over the last several years, Syncrude has been targeting to reclaim approximately 300 hectares per year. This last figure is anticipated to triple as large areas become available for reclamation. By the time these two areas will have been depleted of their resources and subsequently reclaimed, the total land base will be approximately 32,200 hectares. Syncrude still has 3 other leases which are not included in these figures.

The ultimate obligation of the mining companies is to return to the provincial government any mined and disturbed land only after it has been converted to a stable, long-term, self-sustaining, and maintenance-free landscape. Long-term is understood to mean 1,000 years or longer.

Current operations result in large excavated pits and extensive overburden deposits, as well as tailing sands and composite tailing dumps. These dumps, which are in effect massive landforms in a flat landscape, are characterized

Figure 11.22   Aerial photo of equipment removing overburden and excavating oil-sand layer.

as more or less rectangular blocks as much as 5 km long × 3 km wide and 40 meters high with planar, terraced slopes and flat topped plateaus (see Figure 11.23).

Current, conventional practices revealed that planar-slope surfaces, particularly on sandy deposits, even when armored with a surface-water drainage system, can produce severe gullying despite being in a very low precipitation area (see Figure 11.24). This problem was particularly evident after each snow melt. The water from the snow melt created sufficient runoff to cause erosion. Gullies or channel incisions from 4 to 6 feet deep were discovered after each melt, which often removed the vegetative ground cover in addition to damaging or displacing trees and shrubs.

These damages to the man-made dumps required ongoing and expensive repairs and maintenance, and they were in part responsible for preventing such mined areas from being returned to provincial government ownership.

In addition to these erosion and gully problems, other concerns were voiced, for example:

1. The indigenous peoples in the area objected to the stark and radical transformation of their ancestral landscape (see Figure 11.25).
2. The provincial government expressed concern that the dumps were not yet proven to be self-sustaining and instead required continued maintenance. They also commented about the lack of aesthetic considerations when it came to the appearance of the slope faces and landforms.

Figure 11.23 Aerial photo of minetailings and overburden-disposal pile, showing size and extent of artificial landform.

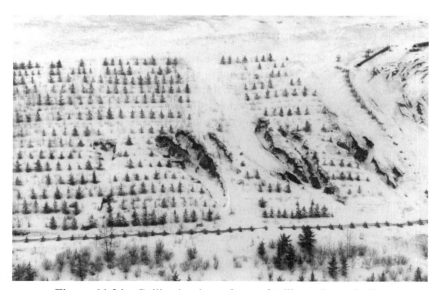

Figure 11.24 Gullies in planar faces of tailings-disposal pile.

Figure 11.25   Aerial view of tailings- and overburden-disposal site showing extent of land disturbance.

3. The forestry department wanted to see more reforestation as an additional requirement.

11.4.6   Repair and Restoration Goals

The basic restoration goal was to dispose of oil-sand tailings and overburden by placing them in self-sustaining, environmentally compatible, and maintenance-free "landforms." The company's closure team recognized that the traditional approach would not meet this established long-term goal.

Initially, these efforts were sometimes hampered by resistance from some sources in the company. While the closure team tried to find creative solutions to their problems, many engineers preferred the simplicity of the traditional, "straight-line" designs.

11.4.7   Treatment Considerations

A variety of treatment alternatives and stabilization techniques were considered initially. These included planting the fine sand disposal piles with native moss similar to that growing on the existing sand dunes, which was effective in stabilizing their movements. Interest was also expressed in the use of *Land-*

form Grading as a way of achieving the repair and restoration goals. To test the validity of landform grading and to convince the operation of its merits, Syncrude opted for a demonstration test. Enhancing a nearly completed fill landform, the demonstration used landform grading principles with a focus on the plateau element. A site known as Southwest Dump 30 was chosen for this purpose.

11.4.8   Selected Demonstrations and Treatments

11.4.8.1   Project No. 1—Landform Demonstration Site No. 1: This demonstration site was a recently completed 50-million cubic meter, 100-meter high disposal pile with a plateau surface of 80 hectares (200 acres). Perimeter slopes were designed from 5H:1V to 10H:1V, with large intermediate terraces. The reason for their relatively gentle inclination was due to the fact that most of the material was uncompacted, with poor foundation conditions, and consisted of weak shale fill. The perimeter slopes were already armored with a gravel layer. Consequently, the focus was not on the slopes but rather the plateau treatment.

Several opportunities existed to give this flat area diverse and interesting natural features and relief. The primary design objectives were as follows:

- To design the plateau of a large tailings "dump" as a show case demonstration project
- To incorporate landforms and vegetation covers that would meet specific requirements established by the provincial government of Alberta and the native and indigenous population

As the design evolved, other stakeholders became involved, expressing their particular interests and specific features they wanted to see incorporated into the design. One proposal was to introduce wood buffalo (bison). To meet this request it was necessary to incorporate buffalo-friendly topography into the overall landscape design. This required the creation of hilltops, which buffalo prefer for resting places and as a place from which to survey the surrounding country side for any approaching predator. Water ponds with a connecting stream were also deemed necessary.

So, in addition to the primary design objectives, the following special features were also incorporated into the overall plan:

- Introduction of 50 bison (wood buffalo) to meet indigenous peoples' desires. To meet this objective, a bison-friendly topography had to be created, namely, open terrain offering visibility, knolls, and high points.
- Forestation measures as requested by provincial government.

- Meandering streambed and ponds (for bison), with ultimate discharge into Base Mine Lake.
- Creation of a vista point for the visiting public.

Alternative grading plans for this disposal facility are shown in Figure 11.26. A conventional plan approach is shown in Figure 11.26a. A grading

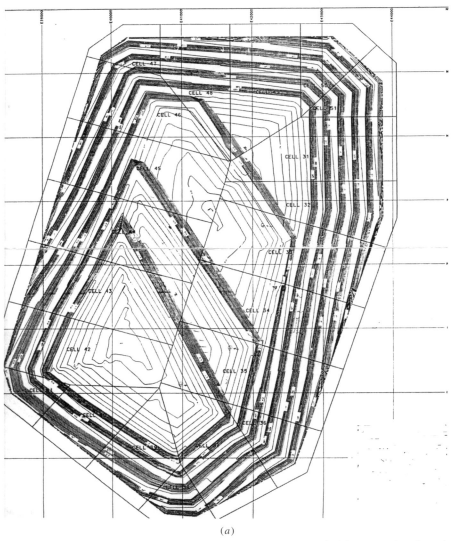

(a)

Figure 11.26 Topographic maps of the top of tailings mound: (a) conventional grading plan, and (b) designed according to landform grading principles.

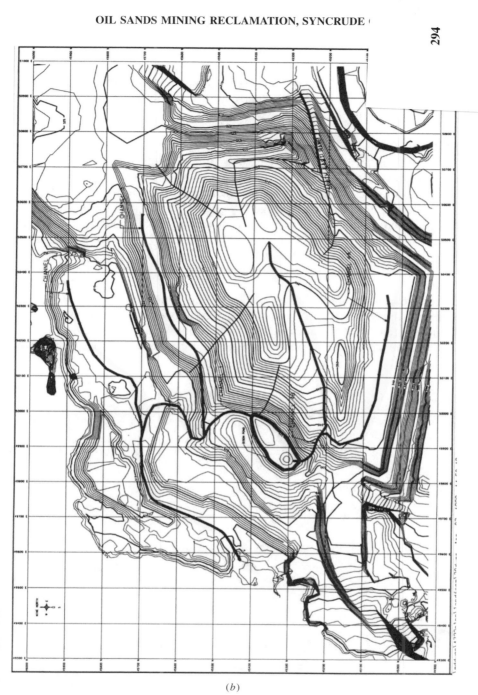

(b)

Figure 11.26 (*Continued*)

plan of the same tailings-disposal facility designed in accordance with the landform grading principles is shown in Figure 11.26*b*. A natural analog for the "plateau top" of a tailings-disposal site is shown in Figure. 11.27. Vegetation is distributed in patterns that conform to the underlying topography as shown in Figure 11.28.

11.4.8.2   *Project No. 2—Reconfiguration of An Existing Tailings Dump Through Landform Grading:* This assignment involved the reconfiguration of the entire, visually exposed frontal area of a large tailings pile. The challenge was to redesign a highly visible, rectangular pile of material with linear slopes in a manner that would make it look like a natural landform. No natural analogs were available to mimic, because the entire landscape for miles around is essentially flat. Consequently, a naturally appearing topographic shape had to be created that would blend into the terrain by following the landform grading and revegetation principles described previously in Chapter 7.

The approach chosen broke up the frontal zone into distinct regions of valleys, ridge lines, and hilltops and divided the zone into smaller tributary drainage areas with meandering valleys to control runoff and avoid erosion while at the same time providing unique landscape opportunities through the

Figure 11.27   Natural analog used as a guide in the designs to visualize the ultimate, mature landscape.

Figure 11.28 Photo showing examples of different tree dispersions in valleys and swales (dark green) and on high ground (light green).

clustering of trees and shrubs in those valleys. After the landform grading plan was prepared, an illustrator was engaged to prepare color renderings of the before and after grading plans that were to be presented to management. Illustrations of his renditions of the existing condition and the landform-grading redesign are shown in Figures 11.29 and 11.30.

Unfortunately, even though approved by management, the landform grading design was not executed, because it was discovered that the original dump slopes had been extensively armored with a gravel surface layer. It would have been cost-prohibitive to remove this surface armor; however, the landform grading plan served as a model for future designs that could be implemented de novo. Furthermore, the landforming plan clearly demonstrated its superiority in terms of ease of revegetation, improved visual appearance, and long-term stability.

11.4.8.3   Project No. 3—"Delandform Grading": This project is presented as a somewhat whimsical example of the early resistance, albeit inadvertent, that landform grading designs sometimes encounter in the field. The example is presented in anecdotal form.

For several days, a survey and construction crew staked and reshaped a previously conventionally constructed linear slope to a landform slope configuration. A day later a new crew arrived to take over the work. Unfortunately, this new crew had not been previously instructed on the matter of landform grading. Upon viewing the strangely shaped slope, the crew chief commented: "The previous crew must have been unaware of the change in

Figure 11.29   Illustrator's rendition of the original conventional grading plan for the tailings disposal.

Figure 11.30   Illustrator's rendering of the landform grading plan.

direction when they built this." So, they proceeded to "straighten things out" and bring the slope back to "normal," that is, "nice" and linear and flat surfaced! Fortunately, the next day the old crew restored the landform design.

11.4.9   Performance Evaluation

Syncrude has found that building these landforms in place of dumps has resulted in several benefits. In the case of the Southwest Dump 30 experience, landform-graded shapes were easier and cheaper to construct than conventional linear and planar disposal piles, in some cases by as much as 20 to 30 percent. These findings were confirmed when Syncrude staff researched and visited some sixty other mine sites in the United States. and Canada. They discovered three other mines—one in Utah, one in British Columbia, and another in Alberta—using methods and techniques that somewhat followed landform grading approaches. These approaches could be described as "roughening the landscape" by creating mounds and irregular topographic shapes. Grading was done more along the lines of contour grading, as described in Chapter 7, Section 7.3, The "Improved" Method.

Revegetation, however, was still applied more along the lines of uniform coverage rather than to explore the opportunities that concave-convex slope elements provide to differentiate plant species and the habitat between the two. Experience also showed in those cases that much less survey control was needed and that the more-detailed parts of the topographic diversity were best achieved by a trained field technician directing experienced and creative equipment operators.

Oil-sands mining faces restrictions in its reclamation efforts, because the entire landscape is underlain by oil sands deposits. Economy of scale has to be utilized when developing creative and more natural landforms for the design of the various overburden, tailing-sands, and composite-tailings deposits and pits.

Furthermore, the excavated soils and the byproducts from the extraction plant, including extremely fine-grained sands, are often not readily conducive to the construction of stable slope and landforms without considerable effort. The perimeter slopes of some tailing deposits have to be laid at angle of 10:1, or flatter, to achieve nonerosive stability.

Today, oil-sand mining companies involved in the extraction process have become very much attuned, along with the regulators, to the need for including the restoration or re-creation of natural landforms and landscapes as an essential element of mining reclamation. Regulations and requirements for closure have changed along with this and it becomes difficult to relinquish the originally reclaimed lands. Pending negotiations should result in an acceptable solution to both parties so that progress and improvements can continue.

11.4.10   Postscript

Nowhere else does the opportunity for landform grading and revegetation exist to the same degree as it does in surface-mining reclamation. The larger the area involved the greater the opportunity to restore real landforms. Nowhere else is the kind of space available to exercise creativity without being restricted by urban boundaries or final-product yield in land development, where every lot and every square foot of buildable land counts.

In the land development environment, all activities are scrutinized by exceedingly sensitive and watchful neighbors who often wield considerable political influence. In spite of such restrictive constraints, landform grading has been applied with great success in hillside housing developments. Think how much easier landform grading applications should be in surface-mining operations, where these kinds of restrictions either do not apply or are not nearly as pronounced.

Surface mining and subsequent reclamation or restoration does have an environmental impact. The final product is mostly the landscape that is left behind as a legacy for future generations. Reclamation based on landforming helps to produce an environmentally friendly and visually attractive landscape.

12 Landforming Projects—Hillside Developments and Mass-Grading Applications

12.1 INTRODUCTION

This chapter describes projects where *landform grading* techniques and procedures were applied to large hillside housing developments and planned communities as well as to the restoration of hillside residential areas impacted by mass grading. Selected case studies include: (1) LA-DWP, Hollywood Hills, a hillside mass grading and canyon restoration project (Los Angeles, California), and (2) Anaheim Hills and Talega master-planned communities (Orange County, California).

12.2 HOLLYWOOD HILLS PROJECT

12.2.1 Type of Project

Landform placement and disposal in adjacent canyon fills of excavated soil for two large underground water reservoirs and rock materials from tunneling operations. Environmental mitigation had to be performed to counter taking 36.4 acres of coastal sage, 1.8 acres of riparian wetlands, and a little over one acre of Army Corps jurisdictional blue-line streambed as part of the overall project.

12.2.2 Location

The site included hill and canyon areas surrounding Lake Hollywood—including a larger, lower, 11-acre water-storage reservoir and a much smaller, upper, water-storage reservoir in the Hollywood Hills (below the famous landmark Hollywood sign). Total site acreage, including watershed area, was 200 acres. The fill disposal, facilities locations, and mitigation sites covered some 40 acres.

12.2.3   Clients

Several clients or stakeholders were involved in this project. They included the following organizations and agencies:

- Hollywood Hills Master Association, a group of seven subassociations surrounding the project site
- City of Los Angeles Department of Water and Power

12.2.4   Projects History

The Department of Water and Power (DWP) of the City of Los Angeles operated an open water-storage reservoir at the site (see Figure 12.1) for decades. Much of the land surrounding the reservoir, including substantial areas of the watershed, was in private ownership.

The City of Los Angeles received a directive from the state's health department (SHD) to provide for covered storage for its water supply at Lake Hollywood. The Department of Water and Power's analysis determined that

Figure 12.1   Aerial view of Hollywood Hills reservoir (Lake Hollywood, California) and surrounding area.

this could be achieved in a cost-effective and environmentally responsive manner through the construction of underground storage tanks. This approach would maintain the current open reservoir as an aesthetic element, provide open-space protection, and assure the same quality of life for the surrounding residential neighborhoods. The most cost-efficient design was to use two 30-million gallon buried concrete tanks. Upon completion in the year 2001, these tanks would become (at that time) the largest prestressed tanks in the USA.

Recognizing that there was no adequate land available under City ownership, the City proceeded to acquire the private property surrounding the reservoir. Initial feasibility studies showed that the excavation for the underground storage tanks and a related tunnel from the tanks to the city's pipelines downstream would generate over one million cubic yards of excavation with no "good" place to put it. The DWP's first proposal was to place this material in several adjacent canyons and valleys. The surrounding community and homeowners immediately voiced strong opposition to this plan. In their opinion, this plan would destroy the natural canyons, ruin all their views, and impact property values negatively.

The DWP then investigated other proposals, such as truck hauling and exporting the excavated material or tunneling and removal by a conveyor belt system. All these alternative proposals suffered from serious environmental impacts and/or excessive costs. The DWP's plan to place the excavated overburden in adjacent canyons was vigorously opposed by homeowners, and attempts to mediate the impasse between the department and the homeowners also failed. It was at this point that homeowners associations learned about the possibility of resolving this impasse by employing landform grading techniques in the event fill was to be placed in the area's canyons. The firm of H. J. Schor Consulting was engaged to assist in developing a disposal plan that would provide an acceptable solution.

In an unusual development, the DWP also contacted H. J. Schor Consulting almost concurrently to assist them in finding a solution that would be acceptable to the community. After several joint work sessions, the community recognized that on-site disposal using landforming was the only viable option. During further negotiations, the City agreed that all land not being used for the Water Quality Improvement Project would be designated as permanent open space and that the lake would be retained with its water-surface elevation maintained by seasonal rainfall and runoff from the watershed.

12.2.5 Earthwork Disposal and Placement Considerations

A study was undertaken of the various canyons and valleys available for disposal of fill. Those that contained substantial tree growth (e.g., sycamores and oaks) and were spring fed were eliminated from consideration. The remaining five canyons and valleys were analyzed for the following attributes:

1. Suitability from the standpoint of accessibility for the grading equipment to prevent scarring of adjacent hillside areas

2. Ability to minimize alteration of existing topography and impact on views over the watershed from surrounding residences

3. Opportunity to preserve clusters of existing trees, in particular the California live oak, and to minimize the area of "taking" of any coastal sage or wetlands, thereby minimizing mitigation costs

4. Fill-holding capacity and suitability for shaping the fills into natural landforms that would harmonize with the adjacent topography

5. Maintenance of *identical* current views, enjoyed by surrounding residences, of the water surface of Lake Hollywood

6. Ability to reestablish a natural hydrological pattern that would supply moisture to the native vegetation, following an interim irrigation system, assisting the initial establishment of the plant palette

Another important consideration was the level of familiarity and comfort with a new grading approach on the part of DWP's engineering, construction, and field personnel. This was essentially a first time experience and challenge for the general contractor to carry out a landform grading design. Even though the general contractor for the entire project, Kiewit Pacific, was a large construction firm with vast experience in a variety of fields, they had never carried out landform grading. It was necessary, therefore, to arrange training sessions first with their management team and then with their field personnel, grade checkers, and equipment operators.

12.2.6   Environmental Mitigation Design Considerations

Some of the obvious canyon locations for fill disposal would involve the taking of natural habitat, coastal sage environment, wetlands, and blue-line streams.

From the start, grading design considerations for the potential fills included not only the desire to replicate natural landforms native to the site but also to develop hydrologic runoff patterns that would provide opportunities for *onsite* mitigation within the body of the newly created slope surfaces, allowing for the establishment and long-term survival of natural habitat regimes. This required the creation of concave slope forms, which would collect, concentrate, and deliver runoff to the locations where it could sustain a given plant concentration and pallet.

The specifications for the revegetation of the slope faces and the final design for the habitat mitigation were developed by Michael Brandman & Associates (MBA) and submitted to the responsible agency, namely, the State of California Department of Fish and Game and the U.S. Army Corps of Engineers, as applicable. This approach was approved by both agencies as acceptable mitigation measures. Fill sites were expected to perform in accordance with accepted standards and reach a level of established maturity guaranteeing long-term success. Ultimately it was possible to accomplish all the mitigation required on-site and on each of the respective landform-graded slopes.

12.2.7 Alternative Grading Studies

A number of alternative grading designs were developed with the objective of achieving a fine balance between the disposal of more than one million cubic yards of soil- and rock-overburden efficiently, while at the same time minimizing the impact on each canyon and valley and preserving the maximum amount of natural open space. Alternatives studies were prepared for consideration by the community and Department of Water and Power.

A consensus eventually emerged after more than 16 alternative grading studies, numerous cross sections and view analyses, and earthwork calculations as well as equally numerous meetings with the community and the DWP. The location and combination of fill sites and quantities allowed for not only the preservation of the primary canyon, fed by a year-round spring and containing dense, native vegetation and a wildlife population, but also the preservation of two additional canyons, sites D and E, which were tributary to the primary canyon.

The final grading design that was accepted is illustrated in Figure 12.2.

12.2.8 Common Design Characteristics of Disposal Fills

The fills were not placed in the side canyons in the traditional manner, namely, as level fills with a sloping 2:1 (H:V) planar triangular-shaped, downstream face. This manner of fill placement and construction has the appearance of a dam-and-impoundment-type structure. Instead, all the fills shared the following design attributes and characteristics:

a. The fills were placed and shaped to replicate the original valley topography but at higher overall elevations. The exception was a single fill that was initially configured as an elevated ridgeline and later reworked into a valley shape.

b. Side slopes of valley fills had concave and convex slope forms incorporated either to simulate the underlying original topography or replicate side-slope forms typical of the area.

c. All valley and slope forms replicated the original slope hydrology so as to maintain runoff patterns and support original plant distributions.

d. Traditional concrete or other hard surface slope-drainage structures were to be avoided or minimized to allow both runoff and infiltration to sustain plant life instead of filling storm-drain pipes.

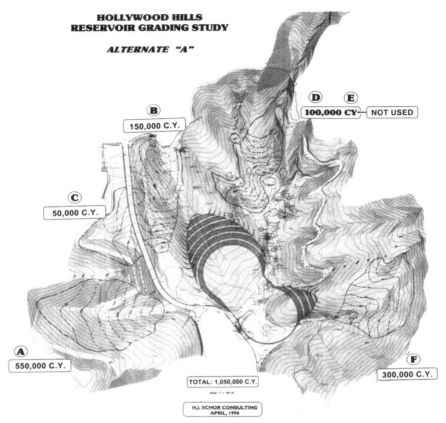

**HOLLYWOOD HILLS
RESERVOIR GRADING STUDY**

ALTERNATE "A"

B
150,000 C.Y.

D **E**
100,000 CY NOT USED

C
50,000 C.Y.

A
550,000 C.Y.

F
300,000 C.Y.

TOTAL: 1,050,000 C.Y.

H.J. SCHOR CONSULTING
APRIL, 1996

Figure 12.2 Accepted grading design showing final locations, import quantities, and basic geometric designs and shapes of the disposal fills.

12.2.9 Final Design Configurations and Locations

A total of four fills were eventually placed. The fills were placed sequentially, starting with Fill F, which was closest to the underground-tank excavation site. The final locations, import quantities, and basic geometric designs and shapes of the disposal fills was noted earlier (see Figure 12.2).

12.2.9.1 Fill F

1. *Site Characteristics*

   This site was the first to be constructed and closest to the proposed underground reservoirs. The site consisted of a broad valley with a fairly uniform south-facing slope and a north face that exhibited a series of concave and convex ridges and tributary valleys. The gross area impacted was 8.2 acres, 6.6 acres of which were coastal sage and

0.9 acres of riparian habitat. A photo of Site F prior to grading is shown in Figure 12.3.

2. *Design Objective and Features*

The objective was to place 350,000 cubic yards of fill in such manner that the original valley shape would be retained and upstream runoff conveyed as before, that is, on the valley floor. The design specified raising the valley floor by some 75 feet and, at the same time, placing the remaining fill up the side slopes. The side slopes were shaped to replicate the underlying natural topography by means of graded convex and concave slope forms, as shown in Figure 12.4. Diagonal ridge features were included as well as elbow-shaped swales (refer to basic landform graded slope forms in Chapter 8). Figure 12.5 shows the revegetation plan for Fill F.

This design not only retained or "re-created" the original valley topography but also maintained the original dendritic drainage and runoff patterns. Runoff from the rear of the fill still flowed on a natural grade out the valley mouth as before. A decision was also made to reinforce the valley floor bottom, where the runoff tended to concentrate, with added plant material. As DWP engineering staff still had concerns regarding erosion, because of lack of experience with this kind of grading, insisting on installing approximately 200 feet of geofabric liner (see

Figure 12.3   Photo of fill site F prior to grading.

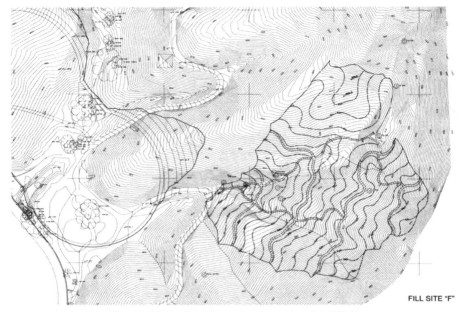

Figure 12.4   Landform grading plan for Fill F.

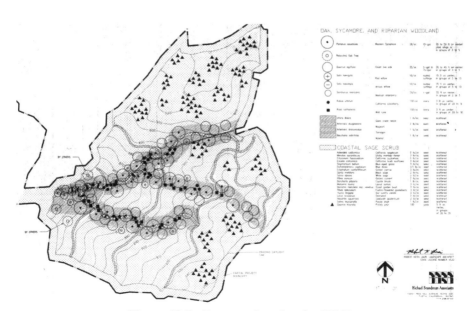

Figure 12.5   Revegetation plan for Fill F.

Figure 12.6) on the steeper terminus of the valley floor and in some side valleys to avoid potential scour erosion from this concentrated flow.

Another design feature, which had to be incorporated upon the insistence of the city's Grading Division (a separate entity), was the placement of benches with concrete terrace drains to intercept slope runoff. The grading division was not convinced initially that a landform graded slope could handle runoff without interception on horizontal terraces or benches with drains. However, the grading division relented and ultimately allowed the use of 50-foot vertical spacing between benches in lieu of the standard 25 feet; but the division still demanded that concrete terrace drains be installed. The increased spacing minimized the artificial or broken appearance of benching on the slope face. To provide some visual mitigation, rock was used to line the concrete drainage ditches; this made them look more like a streambed than a concrete ditch. However, the hardening of the slope-drainage devices had a significant negative effect on the riparian-habitat restoration efforts. Water supply to the riparian borders was carried away instead of infiltrating

Figure 12.6 Photo of geofabric used to line main valley channel or floor bottom.

the soil. Fortunately, the defects of this practice were soon recognized, and the remaining slopes no longer had to be encumbered by terracing and concrete channeling.

3. *Construction*

As the excavation for the two 30-million-gallon tanks commenced, the canyon site was prepared by clearing and grubbing the vegetation cover. All loose and compressible material, such as alluvium and colluvium, was removed and reprocessed as compacted fill to establish a competent base for the incoming, new fill. Prior to fill placement, a subdrain system was installed in the bottom of the main stem canyon and in most of the side valleys that had the potential for future seepage generation and collection.

Initial placement and construction of the landform fill did not start smoothly even though the contractor was knowledgeable about this approach and the field personnel had received sufficient and adequate training. The survey crew had difficulty providing proper and adequate survey control for the grade checker; as a result, the equipment was not placing the fill in the correct manner. The resultant shapes did not appear natural and, instead, began to resemble the old, conventional grading shapes, including an earthen dam structure at the foot of the valley, as shown in Figure 12.7. It was clear that further on-site training on the use of landform grading designs and techniques would be required, along with closer supervision of ongoing operations. Dozers and scrapers were used to remove the "dam" structure and to build up the side

Figure 12.7   Conventional earth dam accidentally under construction at fill site F.

slopes while simultaneously raising the valley floor (see Figure 12.8).

The remainder of the fill was constructed per plan. As time went on, a certain amount of pride was exhibited by all personnel involved as they became accustomed to landforming techniques and as they realized that a landformed fill really did "look" better than fills constructed in

(*a*)

(*b*)

Figure 12.8 Reconstruction at fill site F: (*a*) regrading and reshaping underway and (*b*) river-rock lined terrace drains in place.

the old-fashioned or conventional manner. A series of photos showing the appearance of the landformed fill at Site F, during various stages of construction, are shown in Figure 12.9.

12.2.9.2   Fill B

1. *Site Characteristics and History*

    This fill had quite an unusual history. The original canyon was over 1,000 feet long and 300 feet wide and extended, in almost linear fashion, from Wonderview Drive (at the top) to the new reservoir tank site (at the bottom). Because of the earthwork distribution schedule and per agreement with the community and the DWP this fill had been assigned an importation of 200,000 cubic yards. This amount of fill at this location would have transformed what was formerly a valley, from about 25 feet below Lake Hollywood Drive, into a ridgeline protruding some 40 feet above Lake Hollywood Drive. A total of 6.9 acres was scheduled for regrading, the entire area of which was coastal sage habitat and about 0.1 acre consisting of Federal jurisdiction stream area.

2. *Initial Construction*

    The site required little preparation prior to the placement of imported fill, because practically no compressible material was found after clearing and grubbing. Fill placement proceeded and was completed according to plan.

3. *Redesign Objective and Features*

    Even though the community had been made aware of this fill design, they found it unacceptable after construction; because the new ridge blocked the view of the surrounding hillsides from some residences and, in two cases, also blocked a limited view of the Lake Hollywood water surface. Following extensive negotiating sessions, alternative design plans were prepared for all parties to consider. The alternative ultimately accepted was that the ridge would be reconverted to a "modified" valley. This decision posed several challenges, namely:

    a. How much fill could be accommodated in the original valley and still retain the below street grade valley shape and maintain the predevelopment views for each and every home?

    b. How could the drainage on the fairly steep 7-percent grade valley floor be handled without causing erosion or without resorting to concrete hardening?

    c. What to do with the remaining unplaced yardage?

    The answers developed were as follows:

    a. The final redesign showed that a maximum import of 100,000 cubic yards could remain in the canyon and still allow that yardage to be reconfigured into a valley form.

(a)

(b)

(c)

Figure 12.9   Photos of Fill F during various stages of construction: (a) immediately after grading and hydromulching, (b) one year after construction, and (c) after five years of plant growth and vegetation establishment.

b. To reduce the flow velocity, the flow line of the valley was designed to meander sharply, reducing the overall grade to less than 5 percent. Small natural rock dams 2 to 3 feet in height were constructed in the sharp curves of the meander to prevent erosion and encourage buildup of sediment cones.

c. The meandering valley center made it possible to more extensively landform shape the side slopes by introducing a series of diagonal and curvilinear swales and ridges. The swales were designed to enter at the least erosive location in the main valley channel.

d. A sediment basin area was constructed at the toe of this meandering stream that would capture its flow and feed a newly created wetlands area. These newly constructed wetlands offset the loss of some wetlands removed during excavation of the underground tanks.

e. The 100,000 cubic yards of lost storage capacity was difficult to place as no additional canyons were permitted to be disturbed. Therefore, the Fill A site, the only candidate site large enough to accommodate this extra fill, had to be redesigned to accept it.

4. *Relandforming, from Ridge to Valley Configuration*

As the ground had been previously prepared, the construction process involved removing 100,000 cubic yards of earth and hauling it by scrapers across Lake Hollywood Drive, on a temporary steel bridge, to site A. The residual, or formerly placed fill, was then reshaped and a valley was carved out of the former embankment; finally, the displaced fill was pushed and recompacted uphill, on either side of the newly created valley, as shown in Figure 12.10. An aerial view of the redesigned and graded fill site B is shown in Figure 12.11.

5. *Wetlands Mitigation Fill and Basin*

At the bottom of Fill B, a fill was designed and constructed in the shape of a natural pond with an irregular, natural landform-shaped shoreline. A riser with a downstream-discharge storm drainpipe was installed at a graded low point to provide for overflow release in case runoff and precipitation exceeds the ponds holding capacity. This wetlands-restoration area represented the 0.3 acres of previous taking in a nearby location.

12.2.9.3   Fill C

1. *Site Characteristics*

The smallest of the fill sites, designed to accept only 40,000 cubic yards of imported fill, was less complicated from a design standpoint. The valley was more of a hole leftover after construction of Lake Hollywood Drive on one side and a residential building on the other. It had been used back in the 1930s as a dump site. The total impacted site area was 6 acres. Of this total. 3.3 acres were coastal sage with

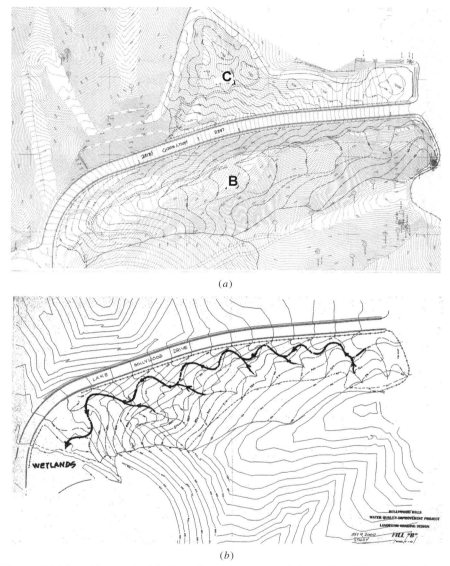

Figure 12.10 Fill B and Fill C grading plans: (*a*) original ridge fill plan for site B and (*b*) revised landformed valley fill grading plan for site B.

no riparian habitat but with 0.3 acres of federal jurisdictional stream-bed area being involved.

2. *Design Objective and Features*

A decision was made to fill the valley slightly above road grade to create a contoured area with an undulating topographic setting. Exten-

Figure 12.11   Aerial view of re-designed and re-graded fill site B shown in center of photo.

sive tree planting created a parklike setting. as the community had desired.

3. *Construction*

Preparation of this site proved to be a sizable challenge. During clearing operations, loose, uncompacted, and wet old fill was uncovered. As excavation and removal of this material progressed, assorted remains of cars from the 1920s and 1930s were discovered along with numerous tires. Clearing operations had revealed an unpermitted old landfill. This material had to be removed and hauled off to a licensed, solid-waste landfill. The rest of the construction proceeded routinely according to plan.

12.2.9.4   Fill A

1. *Site Characteristics*

The side canyon at fill site A was the largest and last of all the imported fill sites. It was located across Lake Hollywood Drive, on the other side of the tank-excavation area (see Figure 12.12). Accordingly, a temporary bridge had to be constructed to provide haul-road access from the excavation area. Total site-impact area was 11.5 acres, of which 9.4 acres consisted of state jurisdictional coastal sage, 0.9 acres of riparian habitat, and 0.45 acres of federal jurisdiction streambed area.

2. *Design Objective and Features*

This site was selected to accept 350,000 cubic yards of fill, according to the initial design agreements. The initial, conventional approach

(a)

(b)

Figure 12.12   Ground photo of valley for Fill A: (a) prior to grading and (b) valley cleared and grubbed, ready to accept import fill.

would have been to place the fill in the usual dam configuration, as shown in Figure 12.13. This design was changed to conform to landform standards, as shown in Figure 12.14. This initial landform design specified a large valley depression in the center of the fill to replicate the original, underlying topography. Initial construction commenced according to this plan. Slope ratios used ranged from 2:1 to 4:1 (H:V).

3. *Initial Construction*

Little preparatory work other than clearing and grubbing was necessary to get the site ready for fill importation and placement. Halfway through this process, however, a redesign was required to increase the holding capacity. This increase was necessitated by the change in Fill

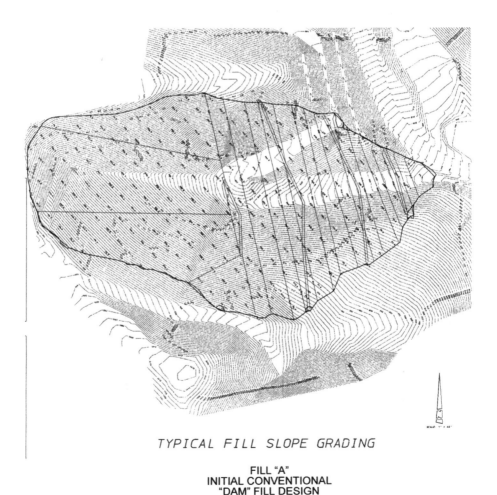

TYPICAL FILL SLOPE GRADING

FILL "A"
INITIAL CONVENTIONAL
"DAM" FILL DESIGN

Figure 12.13 Plan of initial, conventional "dam" design for Fill A.

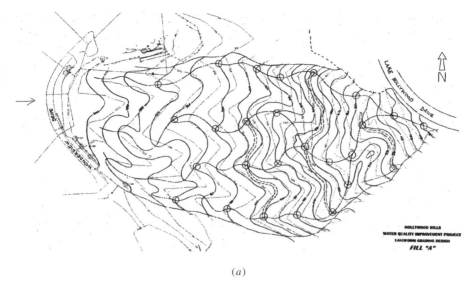

(a)

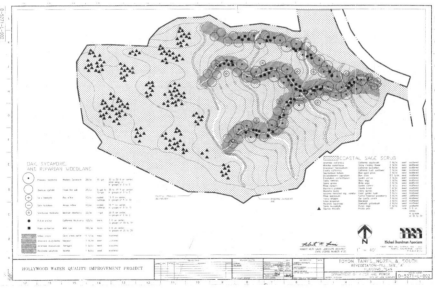

(b)

Figure 12.14   Redesigned plans for Fill A site: (a) topographic redesign to landform standards and (b) revegetation plan for Fill A.

B from a ridgeline to a valley-type fill and the resulting 100,000 cubic yards of additional fill that had to be accommodated elsewhere. An aerial view of Fill A under construction is shown in Figure 12.15.

In this view of Fill A, note how the central portion of the fill is significantly depressed to provide an access road that is less steep for the scrapers hauling the fill import. This fill geometry serves as a small-scale example or model of how larger-scale valley fills, typical of those currently being constructed in the Appalachian coal fields, could be shaped (see Section 7.5.3, Complete or Partial In Situ Landform Restoration). This alternative approach provides for a more gradual transition from a man-made to a natural landform by restoring the valley form instead of creating the typical straight cutoff, dam-type feature. The loss of fill-holding capacity incurred by the depressed center can be compensated (as illustrated in the case of Fill F) by pushing the fill higher up on either side of the valley or even up and above the mountaintop removal plane.

4. *Redesign*

To achieve this new design, grades had to be steepened from the originally planned range of 2:1 to 4:1. The new slopes now ranged from 1½:1 to 3:1 (H:V). The slope surface was also divided into smaller valley and ridge configurations to reduce the size of tributary

Figure 12.15 Aerial view of Fill A during the fill import stage showing the center of the valley depressed for the haul road.

drainage areas for each. This steepening would increase the flow velocities in the concave swales and thereby result in potentially more erosion. Consequently this redesign also required an increase in the number and size of check dams in the flow line of each swale in addition to more intensive planting.

Based on the positive experience with Fill F, the city's Grading Division allowed the elimination of all benching requirements. This also allowed for an increase in the total fill capacity. In addition, it made it possible to create near-level "landing areas" at the slope's midpoint, which allowed runoff to slow down to noneroding velocities.

4. *Reconstruction*

The 100,000 cubic yards of additional material from site B were mainly placed at the upper region of the slope face of Site A after benching into the existing fill to achieve a homogenous mass. While still retaining the overall, initial landform shape, as previously mentioned, the placement resulted in an overall steepening of slope gradients. A further complication ensued when the tunnel-excavated material arrived late in the completion of the A Fill. This more sterile, rocky material ended up being mixed within the upper-slope region or used as top fill instead of being placed deeper within the fill mass. The suitability of this material for plant growth had not been examined previously and created new challenges during the revegetation phase.

Photos illustrating the shape and appearance of Fill A during different stages of placement and construction are shown in Figures 12.16–12.18.

12.2.10   Post-Construction Evaluation and Observations

1. *Department of Water and Power Comments*

The Department of Water and Power (DWP) was very pleased with the results. Landform grading and revegetation allowed them to gain community support and to proceed with their critical project to meet state-mandated deadlines. This approach, furthermore, enabled them to perform all required environmental mitigation on-site, thus saving the cost of property acquisition elsewhere in addition to avoiding logistically more difficult off-site installation and management.

2. *Residents' Comments*

A year after the completion of the first canyon, Fill F, with a substantial vegetative cover in place, a meeting with community representatives and DWP staff took place below Fill A, where site grading was still in process. A homeowner complained about the "terrible" impact of Fill A and pointed across the way to the Fill F location. The homeowner asked why that side canyon or valley site had been preserved intact. He had unwittingly delivered the best compliment and judgment

Figure 12.16 Aerial view of Fill A showing earthmoving and placement.

upon a landform-graded fill slope! The homeowner was unaware, in other words, that 350,000 cubic yards of soil and rock had been placed in the canyon!"

Another, equally interesting experiences arose at the final field inspection meeting between community representatives and DWP staff. It took place at the bottom of the large Fill A site. A resident complained that while the fill and the vegetation in this valley did indeed looked natural, the wildlife had never returned. At this very moment, as eve-

Figure 12.17 Ground view Fill A after completion of fill import.

Figure 12.18   View of completed Fill A one-year later.

ryone turned to look, three deer moved across the filled areas and began
to graze.

3. *Lessons Learned*

Encountered during the construction process, one unplanned and un-
expected problem should be kept in mind in any future projects. The
material excavated from the tunnel was shaley and devoid of any organic compounds. Because of unanticipated scheduling problems, this
material was placed atop the surface of Fill A instead of deep within
the fill. Transplants did not do well, and seeds placed on the fill surface
failed to germinate in spite of repeated applications of hydromulch and
fertilizer as well as regular irrigation. To correct this problem, the DWP
opted to import topsoil and benched a two-foot-thick layer into much
of the slope surface.

4. *Summary of Benefits vis-à-vis Landform Grading and Revegetation*

a. Unanimous acceptance by the community, following protracted bat-
tles and disagreements plus futile attempts to seek other solutions
with the assistance of a professional mediation service. The introduction of landform *grading* concepts initially mitigated and finally
eliminated citizen objections.

b. Cost effective and timely solution to the DWP dilemma to implement
a mandated requirement for the protection of the city's drinking-
water supply.

c. Elimination of any expensive off-site mitigation requirements for the
taking of coastal sage and wetlands habitats. The State of California
Department of Fish and Game accepted these naturally shaped and

revegetated slope areas as acceptable mitigation for the taking of native habitat. All this work was accomplished within the actual construction site (involving no expensive and distant off-site property acquisition) and facilitated the long-term maintenance responsibilities.

d. Reestablishment of the former seasonal runoff pattern and hydrologic conditions on the man-made slopes through *landform grading.*

e. Minimal impact on the quality of life, visual aesthetics, and wildlife habitat. As Fill F proved after five years, this area is hardly distinguishable—to the untrained eye—from the original canyon scenery, both in terms of landforms and vegetative patterns (see Figure 12.19).

f. No negative impact on private property values. Property values in the surrounding area have actually increased significantly.

12.3   ANAHEIM HILLS, CALIFORNIA

12.3.1   Type of Project

A 4,300-acre, master-planned, mixed land-use community in very steep hillside terrain. The project, as finally implemented, included mixed residential

Figure 12.19   Aerial view of Toyon tanks, surrounding area, and Fill Site F five years after construction.

use (primarily single-family detached housing); commercial uses; a public 18-hole golf course owned and operated by the County of Anaheim; a private tennis club; a large, privately operated equestrian facility; parks; schools; and finally eight miles of public trails owned and maintained by Orange County and connected to the county's regional trail system surrounding the project.

12.3.2 Location

The site is in Orange County, approximately 23 miles south of Los Angeles, near the border with Riverside County. Now almost fully developed, approximately two-thirds of the project is situated in the City of Anaheim; the remainder in the City of Orange.

12.3.3 Clients

Anaheim Hills, Inc., the master developer management company (initially a subsidiary of the Robert H. Grant Corp., later of Texaco Inc.).

12.3.4 Jurisdictional Issues

The Anaheim Hills project had to confront difficult jurisdictional challenges. The project fell within or under three separate government or political entities. During the initial planning stages, the County of Orange, the City of Anaheim, and the City of Orange had jurisdiction over various portions of the proposed development.

At the county level, the initial primary issue became the controversial early release from an Agricultural Land Preserve Overlay. This overlay covered about a third of the project site, directly in the path of a rapidly urbanizing area, with most critical utilities already in place for service. On the city side, poorly defined, overlapping, and competing spheres of influence boundaries and jurisdictional issues had to be resolved through the Local Agency Formation Commission.

12.3.5 Site Conditions and Development Constraints

The site is characterized by an east-west trending, largely narrow ridgeline, also known as the Burrell Ridge, with a gentler slope descending to the north and terminating near the east-west running Santa Ana River. This portion was annexed by City of Anaheim. To the south, the ridge exhibits a steeper drop-off, this portion becoming part of the City of Orange Annexation. Geotechnically, the site faced severe challenges from the presence of numerous landslides and faults.

The original master plan showed very intense development patterns, including a proposal for three new lakes with apartments and high-density condominiums clustered around the water features. A water reservoir with a surface area of 80 acres at capacity already existed on the property. This plan was adopted by the City of Anaheim; in anticipation, it would become the major stakeholder because of the size of its projected sphere of influence. As the project entered the more serious engineering stage, and actual geologic investigations took place, it became apparent that the initial master-plan proposal would be unsustainable for a number of reasons:

1. The topography proved to be too steep (over 40 percent of the site was in terrain for slopes 40 percent or steeper), and it would have been cost prohibitive to move the volume of earth per acre needed to generate the necessary level building sites.

2. The geology of the site revealed large existing or suspected landslide masses and fault lines, requiring significant remedial and costly grading operations that the estimated developed land values could not support.

3. Canyon conditions, very deep alluvial deposits and fractured bedrock canyon walls in the location of the dams for the proposed lakes, prevented their construction (an example of when planning and market desires get too far ahead of engineering).

4. The regulatory agency's later desires and the changing market in the direction of lower density required a reevaluation of the land plan.

12.3.6   Initial Development Plan—Community Objections

The revised new master plan that emerged entailed the elimination of the three lakes that would have drowned the canyons. Instead, these areas were proposed to be preserved as parks and open spaces. As grading progressed and early residents moved in, objections were raised about the intensity of the grading, the massive transformation of the landforms, and the lack of any aesthetic consideration. These objections evolved over a period of five years into strong public opposition to the continuation of what was perceived as "raping the hills."

Environmental action groups and antidevelopment forces became more vehement and organized. This opposition manifested itself in the form of frequent letters to local newspapers and strong objections voiced at public hearings before the Planning Commission and City Council. It soon became apparent that the city would be forced into taking action. The opposition clamored for restrictions on grading and density based on the steepness of the topography.

The opposition insisted that the steeper the natural slopes in a given area, the less volume of earth movement per acre and the less density per acre be allowed. They circulated a proposal stating that when natural slopes were 25 percent or steeper, the density was to be reduced to less than one home per acre with little or no grading allowed. Since a little less than 50 percent of the project site exhibited slopes of 40 percent or more, it became apparent that under such restrictions, the master plan for this development, although approved, could never be implemented. And the entire infrastructure, already in place or under construction (streets and highways, sewer, water, storm drains, dry utilities, fire services, police, parks, etc.) and sized for a community of 25,000 residences, would become a wasted expenditure for a population of less than a quarter of that figure.

12.3.7 Adopted Plan—Landform Grading Alternative

Rather than waiting for the City to take more drastic action and thus jeopardize the continuation of the entire project, the principal author, at that time senior vice president in charge of all development functions for the developer, opted to take the initiative. His efforts culminated in the formulation of a *landform grading* and *revegetation* concept. This approach was proposed to the community and the city as a revision to the existing grading ordinance. The new proposal showed clear sensitivity for the land, as it restored or re-created a more natural looking topography and featured an alternative landscaping and revegetation plan in place of the previous rigid, linear, and sterile environment. The city staff was pleased and allowed some initial implementation on an experimental basis. This allowed the decision makers (i.e., planning director, city engineer, planning commission, and city council) and the public to inspect the proposal on the ground.

Following positive reaction from all affected parties and constituencies, the revised grading ordinance proceeded through a brief public hearing process and was formally adopted by the city. The full implementation involved a significant amount of training of planners, civil engineers, geotechnical engineers, landscape architects, grading contractors, and landscape contractors. Fortunately, creative individuals emerged in all these fields, and they met the challenge. The key individual was the grading designer, the individual who "put the pencil on the paper" to execute the concept from which it could be constructed. However, the most important element that drives any innovation is the willingness of all parties to accept and follow through on a design.

With that the Anaheim Hills Planned Community became the pioneering project that first implemented the concept of *landform grading* and *revegetation*. This project became a showcase for others to examine and emulate. Numerous other local and more distant communities, including the City of Los Angeles, sent their representatives to inspect the project and its innovations and to learn from this experience. Visiting professionals from foreign countries were intrigued as well, including representatives from Australia, France, Taiwan, and Japan.

12.3.8 Highlights and Features of Landform Grading Plan

An overhead, aerial photo of the hillside topography prior to and after development are shown in Figures 12.20 and 12.21 respectively. The shape and position of ridges, valleys, and other critical terrain features prior to devel-

Figure 12.20   Overhead, aerial view of terrain prior to development showing position of ridges, valleys, and other terrain features, Anaheim Hills, California.

Figure 12.21   Overhead, aerial view of landform-graded hillside development, showing conformance with underlying terrain features, Anaheim Hills, California.

opment can be seen in Figure 12.20. The same aerial view of the area after landform grading is shown in Figure 12.21. The overall development plan conforms with the underlying terrain features. A bird's-eye, aerial view of the entire Anaheim Hills residential development is shown in Figure 12.22.

The pad shapes and hillside profiles below the pads conform to the landform grading design principles described in Chapter 7. Views of the landform-graded pad areas are shown in Figure 12.23 and Figure 12.24. Essential differences between "conventional" versus "landform" grading approaches to hillside development are contrasted in Figure 12.25. These two grading approaches are also contrasted in the street scenes in Figure 12.26, which illustrates a conventional treatment as opposed to the grading approach used in the Anaheim hills.

12.3.9   Performance Evaluation

The first slopes on this project were graded and revegetated using *landform grading* and *revegetation* principles in 1975. Thus, an over 30-year-long performance record is available. The landform grading and revegetation approach was instrumental in allowing the 4,300-acre hillside master-planned community to be implemented. This approach was well received by the community and by various regulatory agencies. It provided a new and useful marketing tool to the developer and his builders because of its special appeal to buyers and new residents.

Figure 12.22   Bird's-eye, aerial view of completed Anaheim Hills planned community, showing mixed land-use development.

(a)

(b)

Figure 12.23 Views of landform-graded pad area, Anaheim Hills: (a) overhead view and (b) close-up, oblique view.

(a)

(b)

Figure 12.24 Views of building pad areas, Anaheim Hills: (a) graded pad with landform-shaped sides and gradual transition to undisturbed topography and (b) same pad after development showing nonlinear development pattern.

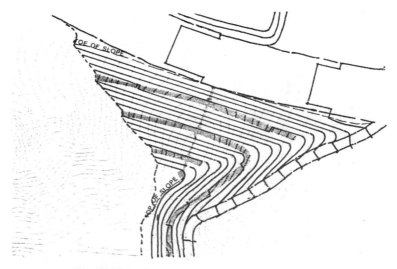

TRADITIONAL GRADING
Straight angular forms and visible bench and down drains
Slope follows roadway with a constant slope ratio

(a)

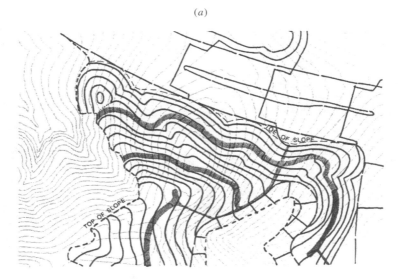

LANDFORM GRADING
Area graded is increased by rounding slopes,creating swales
and knolls and hiding bench and down drains
Hillside is molded to appear natural

(b)

Figure 12.25 Schematic diagram showing basic differences between grading approaches (a) conventional (b) landforming.

(a)

(b)

Figure 12.26 Contrasting street scenes in urban hillside development: (a) conventional practice and (b) landformed in Anaheim Hills.

At no time did the design and construction ever create any structural stability problems, nor did alleged concentrated flows in concave swale areas cause any erosion. Instead, the project became a model and provided a track record for other agencies to feel confident about permitting it in their jurisdiction (see Talega case study below).

12.4 TALEGA, CALIFORNIA

12.4.1 Type of Project

A 3,000-acre master-planned community in steep terrain encumbered by extensive landsliding (approximately 40 percent of the land was impacted by landslides, either active or historic).

12.4.2 Location

At the start of development, the site was partially within the jurisdiction of the Unincorporated Territory of the County of Orange and partially in the City of San Clemente.

12.4.3 Client

Arvida Company. a Florida based master developer of large scale planned communities primarily in Florida and Georgia, had acquired the land for the Talega Project.

12.4.4 Jurisdictional Issues

A particularly unusual component to this development was how the grading logistics were impacted by the political jurisdictional issues.

The master plan for the project had been approved by the City of San Clemente, as all the property lay within the City's sphere of influence, and the City anticipated the annexation by this community upon the completion of development of the property. However, much of the planned development area lay within the County of Orange's jurisdiction.

Because of the topography and the development pattern, almost all of the areas that were low lying and required fill import lay with the City of San Clemente and all the areas that required cutting and export fell within the County of Orange. Each jurisdiction had a very different approach to grading and required that grading under their jurisdiction be performed according to their standards.

12.4.5   Public and Regulatory Agency Responses

The introduction of *landform grading* and *revegetation* into this project produced an interesting series of responses on the part of various public and regulatory agencies and other stakeholders.

12.4.5.1   City of San Clemente's Response: The city had in past years experienced extensive conventional grading, which had very negative impacts on their community, visually, aesthetically, and performance wise. And, under the direction of the city council and the leadership of the community development director, the City insisted that there would be no more "business as usual" when it came to grading. The City took a firm position that there would be no more planar slopes, no more stair-stepping of building pads, resulting in row housing.

The developer and his planners and engineers had made various attempts at modifying their plans, but the changes were not considered to be significant by the City of San Clemente. After almost two years of frustrating planning, replanning, and negotiating sessions, the developer and the city reached an impasse. For the developer this represented a huge financial burden, as the interest clock for the property's purchase price kept ticking away relentlessly, in the form of tens of thousand of dollars a month, as nothing was happening. And the prospect of ever achieving any revenues grew dimmer each day.

Finally the developer approached the city and simply offered: "Just tell us what do you want us to do, and we will do so, so we can move forward."

The City's response was that that was not their obligation, and it was the developer's responsibility to come up with a solution that was going to be acceptable to the City and the community.

As a final resort, the developer relied on a landform grading and revegetation approach to rescue his project. City staff were introduced to the concept with extensive graphic material to provide background. This was well received and considered a positive move. Next, a segment of the proposed development was selected to demonstrate what a change from conventional grading design to the application of landform grading would mean. A planning and engineering demonstration case was developed using the proposed one-mile long major fill slope along the primary entry to the project. This fill slope, some 75 feet in height, was selected to show the change from a conventional to a landform grading design. Along with new grading plans, a three-dimensional model was constructed consisting of two components that could be assembled and disassembled to show what change one would make over the other. For comparison purposes, Figure 12.27 illustrates the original, conventional grading plan, which had been unsuccessfully processed through the City's planning and engineering departments; whereas Figure 12.28 shows the same stretch of roadway redesigned to landform standards.

A site inspection of landform grade areas in Anaheim Hills followed.

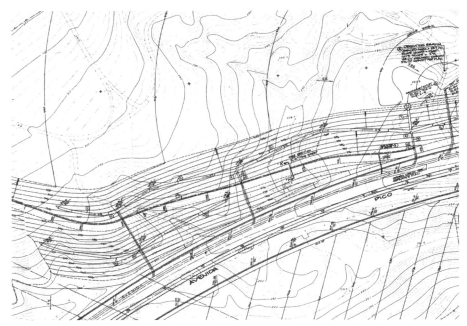

Figure 12.27   Original, conventional grading plan for a section of the primary entry corridor, a six-lane highway (Pico Boulevard) to be entirely constructed as a 75-foot high fill. The plan illustrates the usual application of linear and planar slopes.

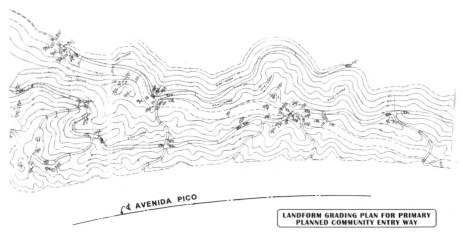

Figure 12.28   The same area redesigned to landform standards.

This visit convinced staff that landform grading and revegetation provided the significant change that it had been looking for, but the staff felt that the proposal must go through the entire public review process. The next steps were public presentations and hearings before the community at large, with a full graphics exhibits, a slide show, a model, and plans. What was presented was a side by side comparison of the original conventional, already fully engineered grading plans for the segment in question, with the landform-redesigned plans, along with conceptual landscape and revegetation plans to demonstrate the differences and benefits derived from this change. The community agreed that these changes met its expectations and enthusiastically embraced them. A public hearing then followed before the planning commission, with a full presentation, and the developer achieved approval for implementation.

The final step was a joint session with the planning commission and the city council, and unanimous approval was obtained from all with the stipulation that all future grading and landscape plans for the Talega Project would incorporate the Landform Grading and Revegetation Standards.

12.4.5.2   The County of Orange's Position: The County had a very rigid ordinance requiring strict application of conventional design, grading, and landscaping principles. The county had probably the most experience with hillside grading and development, and their standards were purely based on securing a safe and stable development without any consideration to aesthetics. This was in the public's view a major detriment to allowing the grading to occur within the county's jurisdiction, so close to the city, which would ultimately be incorporated into the City of San Clemente. Therefore, it became necessary to convince, also, the County of the merits of landform grading and revegetation and to allow sufficient modification to the County's standards to implement such. This proved initially far more difficult than the process at City level, as this was a much larger agency that was firmly entrenched in bureaucratic processes, deviating from which would take considerable effort.

To facilitate the processing of plans through approval and supervision of the grading operations, the County and the developer, in a joint effort, developed detailed standards and guidelines for the County's plan checkers and field inspectors. These functioned well through the life of the project. Again, civil engineers, geotechnical engineers, and landscape architects "stepped up to the plate," driven by the unacceptable consequences of a "no-option" position, to perform their tasks and to achieve the desired results.

12.4.5.3   Final Resolution and Agreement Among Stakeholders: In this case, the decision to implement landform grading and revegetation was neither a "voluntary" one nor a desire to satisfy the stakeholders' own desires; rather, the decision was motivated by the strong desire and determination of a community and its professional and political representatives. The final implementer, the grading contractor, after some disappointing trial runs, brought its best equipment operators and grade checker to the project and changed the internal procedure and "voila!" Landform grading emerged on a grand scale. The County was so pleased with the results that the developer received a letter of commendation for his grading, something unheard of before. Typically, the developer and the agency would be accused of "raping the hills" again.

The city was even more exited and pleased that a dedication ceremony with members of the city council, planning commission, staff, and the news media was held at the site of the completed mile-long highway slope, where the concept was first implemented, with the mayor stating that "this is what my community has been looking for."

12.4.6   Highlights and Features of Project

The scale and visual impact of the completed landform-shaped slope, along the primary entryway to the project, are illustrated in Figure 12.29. Two close-up views are shown in Figures 12.30*a* and 12.30*b*. Figure *a* shows the overall size and shape; Figure *b* illustrates the concept of breaking the slope surface into smaller tributary drainage areas to minimize erosion.

The landform-shaped slope also offered an ideal platform for revegetation that differed from a slope with the customary landscaping treatment. The landformed slope provided opportunities for clustering of trees and shrubs in

Figure 12.29   Completed landform-graded primary entry slope to project.

(a)

(b)

Figure 12.30   Detailed photos of slope forms along primary entry: (a) distant view of landformed slope and (b) close-up view showing slope-drainage system.

the swales or valleys as opposed to arbitrary scattering or uniform coverage. The revegetation plan that was developed for the landformed slope is shown in Figure 12.31. A photo of the revegetation work in progress is shown in Figure 12.32.

The wide variety of slope forms and shapes allowed for creative design of drainage structures and their placement in a more natural manner. The use of river rock to line the drainage channels further enhanced their visual appearance. Figure 12.33*a* shows a drainage channel under construction; Figure 12.33*b* depicts the finished installation. Note how the drainage structure "wraps" itself around the slope forms.

Another challenge encountered within the project was to shape a 1,100-foot long × 120-foot high cut slope into a natural landform. Adverse bedding conditions were discovered when this slope was being cut. This finding required that the slope be redesigned and reconstructed using a buttress fill. Figure 12.34*a* depicts the grading plan for this slope. Note the diagonal and curvilinear placement of down-drains on the slope.

To achieve adequate compaction on the slope face, the entire slope was overfilled by an equipment width (of a scraper and dozer) and then compacted. The fill was then cut back to finish grade while landform shaping it, at the same time, as shown in Figure 12.34*b*. Once ground cover was established on the surface, it was undistinguishable from the adjoining natural slope. The finished and planted, 120-foot high buttressed cut slope is shown in Figure 12.35. A later survey showed that people driving by this slope considered it to be an undisturbed, natural slope.

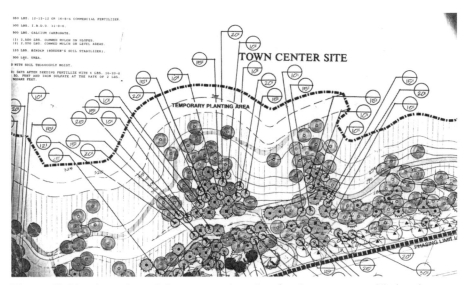

Figure 12.31 A section of the revegetation plan for the same area. Notice the concentration of plants and trees in concave slope forms.

Figure 12.32 Landscaping work-in-progress, using the revegetation technique.

12.4.7   Conclusions

12.4.6.1   Operational Findings: This project showed that while cut-and-fill slopes can readily be landformed shaped, the construction of buttresses or stabilization fills on false cuts could not be performed as a single operation. In other words, slopes could not be constructed from the ground up in land-form shapes and still achieve 90 percent compaction through track rolling. These slopes, in all cases, had to be overfilled and then cut-and-shaped back, which is often the preferred method even for conventional grading.

12.4.6.2   Economic Considerations: The civil engineering cost, because of the training process, did incur additional costs in the neighborhood of 1 per-cent, which included design and staking.

Geotechnical engineering reviews, analysis, recommendations, and field supervision and inspection did not generate any measurable additional expenses.

The grading contractor found that cut-and-fill operations could be performed with conventional equipment using largely conventional logistics and techniques—i.e., compaction within the same cycle times (earth movement times from cut to fill placement) as the "mass X" (see Figure 9.18), the bulk volume is still moved in the conventional way.

The only significant additional costs were incurred in the finishing of the final slope face features. When the final accounting was made, it was found that from the total cost of this grading project of $25 million, less than 2 percent was attributable to the implementation of landform grading.

(*a*)

(*b*)

Figure 12.33   Photos of down-drain installation in the landformed slope: (*a*) during installation and (*b*) finished installation.

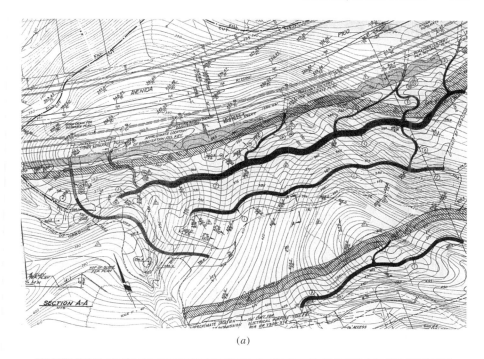

(a)

(b)

Figure 12.34 Landform graded and buttressed 120-foot high cut slope along highway: (*a*) landform grading plans for cut slope and (*b*) Freshly graded slope according to plans.

Figure 12.35   Finished and planted landform-graded and buttressed cut slope.

The implementation of the revegetation plan required no additional cost—neither for design nor planting nor for the installation of the irrigation system. No additional plant material was required, just an adjustment as to "where things go" and specific directions in the field so that workers used to the equal spacing concept would understand the new planting concept and purpose of clustering.

APPENDIX
Survey Staking for Landform Slope Construction

The development of appropriate survey-construction-staking techniques is essential for successful implementation of landform-grading construction. Traditional cut-and-fill grading produces contours that are either straight (tangent) or follow flowing, generally long, radii curves; much like those along roadway centerlines. By contrast, landform grading produces short radii contours or contours that swirl and dip with considerably more curves or elements than in traditional grading or even contour grading.

The purpose of landform grading is to produce earthwork contours that are pleasing to the eye and that mimic naturally occurring topography. The challenge for the surveyor is to relate the many diverse elements of landform-designed contours to the grading contractor in a manner that will enable this type of slope construction to proceed accurately and efficiently.

It is important to note that, in Southern California, landform grading is most frequently a "fill"-grading operation. Should this type of contouring be required in a naturally occurring cut slope requiring stabilization, a back cut is graded an equipment width or so (about 15 feet) behind the plane of the designed slope, and grading then occurs as a fill operation. Grade stakes are thus mostly fill stakes. With this in mind, the slope itself will generally be overfilled, then trimmed to produce the designed slope. The grade checker and equipment operator will need to know "where" on the slope all landform elements occur, even those between surveyed stations. One might assume that, given the detailed nature of landform design, a rather prodigious amount of staking, with a correspondingly high cost, must be performed to keep control of slope construction. Fortunately, this does not have to be the case. In fact, survey-staking costs can be very much the same for landform grading as for traditional grading.

Traditional grading requires the setting of slope stakes at approximate 50-foot intervals along the slope. Slope returns are staked generally at one-half delta or one-quarter delta intervals, depending on slope radii. Cuts or fills are marked on the lath with the "story" of the slope to be constructed. A large fill slope might be staked thusly: fill 1 foot out 10 feet to the toe, then fill 30 feet at 2 to 1, out 60 feet, then level 8 feet, then fill 30 feet at 2 to 1 out 60 feet, then level 8 feet then fill 21 feet at 2 to 1 out 42 feet to the top elevation of 108 feet. The abbreviation on the stake would read as follows:

F-1^0 × 10^0 Toe / / F-30^0 @ $2:1$ × 60.0 / / Level 8^0 / /
F-30^0 @ $2:1$ × 60^0 / / Level 8^0 / / F-21 @ $2:1$ × 42^0 Top (108^0).

The key to landform-grade staking is always to know where are the major contours. Instead of slope ratios and outs, landform-slope staking will show the grade checker fills and outs to (typically) even 5-foot contours. The grade checker must be competent at the craft. The surveyor need only to set slope stakes along 50-foot stations (or so) to accomplish this, but the grade checker must know where the slope-stake lines cross the contour lines to determine the location of landform elements, which fall between the surveyed lines. To do this, the grade checker needs a plan.

When establishing the location of the slope stakes, the surveyor takes the plan and draws lines perpendicular to the slope along each line intended to be controlled. If the slope falls along a roadway, the stake lines are at even 50-foot stations along the roadway. So far, this is the same as traditional slope staking. The surveyor then scales the distance between each even 5-foot contour and writes it in the plan along the intended slope-control line. The top and toe of the slope should, of course, match all criteria for proper location and elevation; as shown on the plan. As with normal slope staking, the stakes are then set in the field at an agreed upon offset.

The actual blue-line plan that the surveyor uses for the staking set up is then reproduced and a copy given to the grade checker, along with a copy of the survey field notes. The stakes and the notes will read distances and outs for major contours.

For example, a typical landform slope stake may read: fill 1 foot out 10 feet to toe, then fill 3 feet out 6 feet to contour 95 then fill 5 feet out 13 feet to contour 100, then fill 5 feet out 7 feet to contour 10^5 and so on to the top. The abbreviation would look like this:

F-1^0 × 10^0 Toe / / F-3^0 × 6^0 Cont. 95^0 / / F-5^0 × 13^0 Cont. 100 / /
F-5^0 × 7^0 Cont. 105^0 and so forth.

The grade checker proceeds to construct the slope, referring to both the survey stakes and staking plan, noting all elements of landform contouring occurring between the staked 50-foot stations. The grade checker simple prisms these contours in, using a right-angle mirror and a hand level. At no time is the grade checker ever more than 2½ vertical feet or more than 25 horizontal feet from a controlled contour line. In addition, landform elements occurring on a slope are primarily aesthetic and thus a greater degree of tolerance for the placing of these elements can be allowed than are allowed along traditional slope planes.

The surveyor can also digitize the major contours from a grading plan and upload these points into one of many field-data collectors presently in use in the survey industry. Slope staking can then be accomplished from an eccentric point, and slopes under construction can be verified easily and accurately.

Grading acceptance of landform grading by the surveyor is generally limited. The surveyor verifies the location and elevation of all tops and toes but does not attempt acceptance of the slope plane in between. This acceptance of the actual landform elements is performed by the jurisdictional agency and developer. Aerial topography can be performed at the conclusion of grading should a detailed record of existing contours be desired.

In summary, slope staking for landform grading can be performed with none, or only very minor, cost increases over traditional slope staking. Teamwork and communication between the grading contractor and survey party chief will assure a high-quality product at a very reasonable cost. As landform grading gains acceptance, the level of competence in staking and construction techniques will rise to meet the developers' demands, keeping costs as competitive as possible.

This appendix was prepared with the assistance of Michael R. Miller, PLS, Vice President Surveying, RBF Consulting Engineers, Irvine, California.

Index